ERCOFTAC SERIES

VOLUME 3

Series Editors

P. Hutchinson, *Chairman ERCOFTAC,*
Cranfield University, Bedford, UK

W. Rodi, *Chairman ERCOFTAC Scientific Programme Committee,*
Universität Karlsruhe, Karlsruhe, Germany

Aims and Scope of the Series

ERCOFTAC (European Research Community on Flow, Turbulence and Combustion) was founded as an international association with scientific objectives in 1988. ERCOFTAC strongly promotes joint efforts of European research institutes and industries that are active in the field of flow, turbulence and combustion, in order to enhance the exchange of technical and scientific information on fundamental and applied research and design. Each year, ERCOFTAC organizes several meetings in the form of workshops, conferences and summerschools, where ERCOFTAC members and other researchers meet and exchange information.

The ERCOFTAC Series will publish the proceedings of ERCOFTAC meetings, which cover all aspects of fluid mechanics. The series will comprise proceedings of conferences and workshops, and of textbooks presenting the material taught at summerschools.

The series covers the entire domain of fluid mechanics, which includes physical modelling, computational fluid dynamics including grid generation and turbulence modelling, measuring-techniques, flow visualization as applied to industrial flows, aerodynamics, combustion, geophysical and environmental flows, hydraulics, multi-phase flows, non-Newtonian flows, astrophysical flows, laminar, turbulent and transitional flows.

High Performance Computing in Fluid Dynamics

Proceedings of the Summerschool on
High Performance Computing in Fluid Dynamics
held at Delft University of Technology, The Netherlands,
June 24–28 1996

Edited by

P. WESSELING

Faculty of Technical Mathematics and Informatics,
Delft University of Technology,
The Netherlands

Held under the auspices of and with support of
ERCOFTAC (European Research Community on Flow, Turbulence and Combustion)
and the J.M. Burgers Center (Graduate School for Fluid Dynamics, The Netherlands).

Kluwer Academic Publishers
Dordrecht / Boston / London

A C.I.P. Catalogue record for this book is available from the Library of Congress.

Published by Kluwer Academic Publishers,
P.O. Box 17, 3300 AA Dordrecht, The Netherlands.

Kluwer Academic Publishers incorporates
the publishing programmes of
D. Reidel, Martinus Nijhoff, Dr W. Junk and MTP Press.

Sold and distributed in the U.S.A. and Canada
by Kluwer Academic Publishers,
101 Philip Drive, Norwell, MA 02061, U.S.A.

In all other countries, sold and distributed
by Kluwer Academic Publishers Group,
P.O. Box 322, 3300 AH Dordrecht, The Netherlands.

ISBN-13: 978-94-010-6606-8   e-ISBN-13: 978-94-009-0271-8
DOI: 10.1007/978-94-009-0271-8
Printed on acid-free paper

TABLE OF CONTENTS

Preface

This book contains the course notes of the Summerschool on High Performance Computing in Fluid Dynamics, held at the Delft University of Technology, June 24-28, 1996, under the auspices of and with support of ERCOFTAC (European Research Community on Flow, Turbulence and Combustion, WWW: http://imhefwww.epfl.ch/lmf/ERCOFTAC) and the J.M. Burgers Center (Graduate School for Fluid Dynamics, The Netherlands, WWW: http://tnj.phys.tue.nl/fdl/Burgers.html).

In addition to the material presented in the present volume, lectures were given by J. Häuser on "A parallel Newton-GMRES method for the 3D Navier-Stokes equations in complex geometries". The material written down by F.-S. Lien was presented by M.A. Leschziner.

This book is addressed to students on a graduate level and researchers in industry engaged in scientific computing, who have little or no experience with high performance computing, but who want to learn more, and/or want to port their code to parallel platforms. Applications in computational fluid dynamics are emphasized.

The lectures presented here deal to a large extent with algorithmic, programming and implementation issues, as well as experiences gained so far on parallel platforms. Attention is also given to mathematical aspects, notably domain decomposition and scalable algorithms. Computer science basics are not emphasized. Topics considered are: basic concepts of parallel computers, parallelization strategies, programming aspects, parallel algorithms, applications in computational fluid dynamics, the present hardware situation and developments to be expected. This reflects the order in which the material is presented. There are ample references to the current literature, so that the present work is a good starting point for those who want to enter the field of high performance computing, especially if applications in fluid dynamics are envisaged.

P.Wesseling   Delft, March 1996

List of speakers

M.J. Daydé
ENSEEIHT-IRIT
2 Rue Camichel, Toulouse, France

D.R. Emerson
Computational Engineering Group, Daresbury Laboratory
Keckwick Lane, Daresbury
Warrington WA4 4AD, Cheshire, United Kingdom

J. Häuser, CLE Center of Logistics and Expert Systems
Karl-Scharfenberg-Strasse 55/57, 38229 Salzgitter, Germany

C. Lacor
Department of Fluid Mechanics, Free University Brussels
Pleinlaan 2, 1050 Brussels, Belgium

M.A. Leschziner
The University of Manchester Institute of Science and Technology
P.O. Box 88, Manchester M60 1QD, United Kingdom

P.H. Michielse
Hewlett Packard Company / Convex Computer BV
Europalaan 514, 3526 KS Utrecht, The Netherlands

M. Streng
Department of Applied Mathematics, University of Twente
P.O. Box 217, 7500 AE Enschede, The Netherlands

A.E.P. Veldman
Department of Mathematics, University of Groningen
P.O. Box 800, 9700 AV Groningen, The Netherlands

H.A. van der Vorst
Mathmatical Institute, University of Utrecht
P.O. Box 80.0101, 3508 TA Utrecht, The Netherlands

Editor
P.Wesseling
Department of Technical Mathematics and Informatics, Delft University of Technology
Mekelweg 4, 2628 CD, Delft, The Netherlands

INTRODUCTION TO PARALLEL COMPUTERS: ARCHITECTURE AND ALGORITHMS

D. R. EMERSON
Computational Engineering Group
Daresbury Laboratory
Keckwick Lane
Daresbury
Warrington WA4 4AD
Cheshire
United Kingdom

1. Introduction

The term "parallel processing" refers to performing computational tasks concurrently i.e. at the same time. Each individual task can be either identical, such as adding two elements of an array, or unique. It should be made clear at this stage that parallel processing is not a new concept. Indeed, it has been around in practice for several decades and its concept has been understood for over 150 years (Menabrea, 1842). However, it is only since 1985 that commercially available machines have been in use outside of purely academic or research establishments. Furthermore, it is now widely recognised that only parallel processing offers the potential of solving very large scale scientific problems, such as those encountered in Computational Fluid Dynamics (CFD), Computational ElectroMagnetics (CEM), Quantum Chemistry, Quantum ChromoDynamics (QCD), and many other computationally demanding fields, in an "acceptable" time and at a "reasonable" cost. Of course, what is acceptable or reasonable will very much depend upon each individual's requirements.

One of the significant differences between conventional (sequential) programming and parallel processing is that it is necessary to have an understanding of the underlying computer architecture and the numerical algorithm and, particularly, what effect one has upon the other for efficient operation. This is an extremely important feature in order to extract a significant fraction of the peak performance from the machine, a task that is frequently difficult to achieve in practice. It therefore begs the question: Why do we need to use parallel processing? There are several answers to this question and each answer depends upon your own perspective. The "Grand Challenge" topics listed above can clearly only be done on the fastest computers with the largest memory. However, many other applications, both commercial and scientific, can greatly benefit from parallel processing in several ways, such as: (i) reducing the time a particular computation takes and therefore improving productivity and throughput; (ii) providing access to more processors (and therefore more memory) so that bigger problems can be tackled in the same total time; and (iii) providing a very competitive price/performance. The foregoing list contains the most cited examples in support of parallel processing

1

P. Wesseling (ed.), High Performance Computing in Fluid Dynamics, 1–42.
© 1996 *Kluwer Academic Publishers.*

but it is not exhaustive and other equally valid views can be included. Indeed, there is a fundamental limit to the speed of a processor which cannot be overcome by current scientific knowledge and this is dealt with in section 3.

2. Taxonomy and Classification of Computers

Flynn (1972) introduced a classification of computer systems that is now widely used by the computing community. He divided machines into 4 categories based on how the machine relates its instructions to the data being processed. The categories are:

1. SISD - Single Instruction stream/Single Data stream
2. SIMD - Single Instruction stream/Multiple Data stream
3. MISD - Multiple Instruction stream/Single Data stream
4. MIMD - Multiple Instruction stream/Multiple Data stream

Table 1 Relationship between instruction and data streams
using Flynn's taxonomy

Flynn's Taxonomy		Number of Data Streams	
		Single	Multiple
Number of Instruction Streams	Single	SISD	SIMD
	Multiple	MISD	MIMD

The term *stream* relates to the sequence of data or instructions as seen by the machine during execution of a program. An instruction stream is a sequence of instructions as executed by the machine and a data stream is a sequence of data including input, partial or temporary results, called for by the instruction stream. The SISD machine represents the von Neumann model and is representative of most serial scalar computers. Parallelism can be introduced by operating on multiple data but using the same instruction. This is the SIMD model which corresponds to either an array processor (such as the ILLIAC IV, CM2, ICL DAP) or a pipeline architecture (such as the CRAY 1). The general view regarding the MISD classification is that this machine does not exist and is only included for completeness. However, there is a view that perhaps the pipelined vector processor belongs in this category and not under the SIMD category. The final category is the MIMD class of machine. This includes all forms of multiprocessor configurations (e.g. CRAY X-MP, Intel iPSC/860). These relationships are further indicated in Table 1. It is easy to see that the foregoing classification scheme is generally too broad to distinguish between many very different architectures and, in practice, many computers are hybrids (e.g. the CRAY X-MP is a MIMD architecture with each processor using pipelining i.e. a SIMD operation). It is therefore not surprising that several other classification schemes exist. For more information see, for example, Kuck (1978) or Gajski and Pier (1985). Although Flynn's taxonomy has several weak areas it is still the most widely used classification system and suitable for most purposes.

3. A Brief History of Computers and Parallel Computing

Whilst it is beyond the scope of this book to give a detailed account of computing through the ages, it is illuminating to briefly review the major advances made since the 1950s when computers first became available. The earliest work on computers was pioneered by Babbage (1821) (see Randell, 1982) on a machine he called a Difference Engine. Babbage estimated this machine to be capable of performing one addition or subtraction in one second. It was also estimated that a multiply or divide would take one minute. The simultaneous computation of numbers to improve the performance of the machine was also envisaged by Babbage but this could not be achieved with the technology of the day. Indeed, apart from a small working model, Babbage did not complete the full scale machine, whose technological requirements were beyond the capabilities of the era. However, other people did successfully build upon his work and the earliest reference to employing parallelism as a means of speeding up the computational process is by Menabrea (1842) who recognised that for identical computations, such as those found in the formation of numerical tables, a machine could be used to produce several results at the same time.

The present day history of computers really begins between 1938–1953 when valve technology was employed. These computers are known as the *first generation* (Hockney and Jesshope, 1988, Hwang and Briggs, 1985). Electronic valves were used as switching components and had gate delay times of about $1\mu s$ (10^{-6} seconds). The gate delay time is how long it took for a signal to travel from the input of one logic gate to the next logic gate (Turn, 1974). The first stored program computers were the Electronic Discrete Variable Arithmetic Computer (EDVAC), at the University of Pennsylvania, and the Electronic Delay Storage Automatic Calculator (EDSAC), at the University of Cambridge (Randell, 1982). The latter machine, the EDSAC, is generally regarded as being the first completed version of a stored program computer. After beginning work in early 1947, the first program ran in May, 1949. One of the lasting benefits of this early work was the realisation that the computer could be used to produce its own programs and thereby the foundations of work on assemblers, compilers and operating systems was developed.

Semiconductor technology was introduced circa 1960, using discrete germanium transistors with gate delay times of approximately $0.3\mu s$ and formed the basis of the *second generation* machines, which include the IBM 7090. The second generation of computers lasted from 1952–1963. The transistor was invented in 1948 and high level languages became available in the late 1950s, including FORTRAN (FORmula TRANslation) in 1956 and ALGOL (ALGOrithmic Language) in 1960. The third generation (1962–1975) saw the advent of solid state memory. By 1965, Small Scale Integration (SSI) was being employed with a few gates per chip and gate delay times of about 10ns (1ns = 10^{-9} seconds). Improvements in technology led to a reduction in the gate delay time to less than 1ns by 1975. Medium Scale Integration (MSI) brought further advances. Large Scale Integration (LSI) heralded the *fourth generation* of computers and Very Large Scale Integration (VLSI) is now the norm and was used by the *fifth generation* of computers looking at Artificial Intelligence (AI).

From the foregoing, the component speed has increased by a factor of 1000 from the year 1950 to 1975. However, the processing speed, as measured by the time it takes to perform a floating point operation, had increased by approximately 10^5. This substantial increase had been brought about by the introduction of parallelism at the architectural level. It is also worth noting that the CRAY 1 was delivered to the Los Alamos Scientific Laboratory in 1976. By employing pipelining techniques, the CRAY 1 had a peak performance of 160 Mflop/s (160 million floating point operations per second). Another major success around this period was the ILLIAC IV. The work on this machine began in the mid 1960s and followed on from previous pioneering work performed on the SOLOMON (Simultaneous Operation Linked Ordinal MOdular Network) computer. The ILLIAC IV became fully operational during November, 1975 (Hord, 1982). It was the first large scale array processor and it is widely regarded as the first supercomputer. Its development brought many advances to computing technology, including contributions to logic circuitry, the use of Emitter Coupled Logic (ECL) and it was the first large scale machine to use all semiconductor memory. The ILLIAC

Figure 1 Typical SIMD Architecture Arrangement

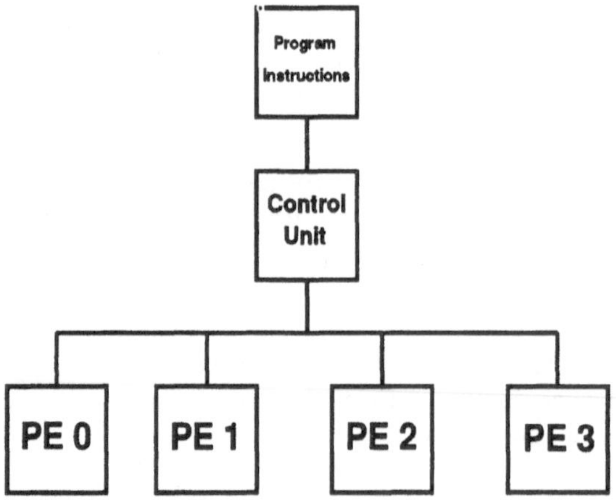

IV was developed because the scientific community wanted faster machines and this demand for high speed computing from sequential machines was beginning to hit upon the ultimate barrier - the speed of light. This is the theoretical limit at which a signal can propagate through a conductor and it therefore imposes a physical barrier that cannot be overcome. At this speed, the information can travel approximately 0.3m in one nanosecond. In reality, however, the speed of propagation is somewhat less than the speed of light and the distance is therefore less. Due to this fundamental limitation it is necessary to minimise the distance a signal has to travel but even though significant improvements have been made in this direction, this approach is eventually limited by quantum mechanics. To overcome the limit imposed by the speed of light, the designers of the ILLIAC IV opted for a parallel approach which employed 64 processing elements (PEs), each with its own memory (in this case, 2048 words of 64 bit length) and all simultaneously operating the same instruction on different data. A single control unit is used to broadcast the instruction to each PE and figure 1 illustrates the array processor architectural layout. It is also worth noting that both the CRAY 1 and the ILLIAC IV used a SIMD approach to try to overcome the physical limit of the speed of light.

In the late 1970s and early 1980s, distributed memory architectures were investigated by a wide range of groups. However, it was not until the early to mid 1980s that machines containing more than one processor and conforming to the

Multiple Instruction stream Multiple Data stream (MIMD) classification were available commercially. At this stage, it is worth distinguishing between the two major MIMD categories, namely Shared Memory MIMD (SM-MIMD) and Distributed Memory MIMD (DM-MIMD) computers. The SM-MIMD group are considered to be tightly coupled, whereby the instruction streams can be programmed to work together on the solution of a single problem, whilst the DM-MIMD are regarded as loosely coupled and employ the relatively slow message passing approach. The CRAY X-MP, containing essentially 2 CRAY 1 processors, was announced in 1982 and was the first multi-processor to be shipped (Dongarra et al, 1991) and by 1984 a 4 processor version was also available. Each machine used pipelining to achieve very high performance and the machines had a peak performance of 420 and 840 Mflop/s, respectively. In 1985, the Intel iPSC (Personal SuperComputer) hypercube, the commercial variant of the Cosmic Cube, entered the supercomputing market in sizes of 2^5, 2^6 and 2^7 and the largest system was rated at 8 Mflop/s. However, by 1986 Intel were offering improved node performance by the inclusion of a pipelined vector processor. This increased the peak performance of each node to 20Mflop/s at the expense of reducing the total number of nodes to 64. However, the overall peak performance had now increased to approximately 430 Mflop/s in 64 bit arithmetic.

It was previously stated that the component speed had shown a thousand fold increase between 1950 and 1975: Has there been a similar dramatic increase since 1975? Figure 2 illustrates many of the well known computer systems and attempts to answer this question. From figure 2, it is evident that it has not. Since 1976 (the

Figure 2 Clock cycle time (in nanoseconds) since 1975

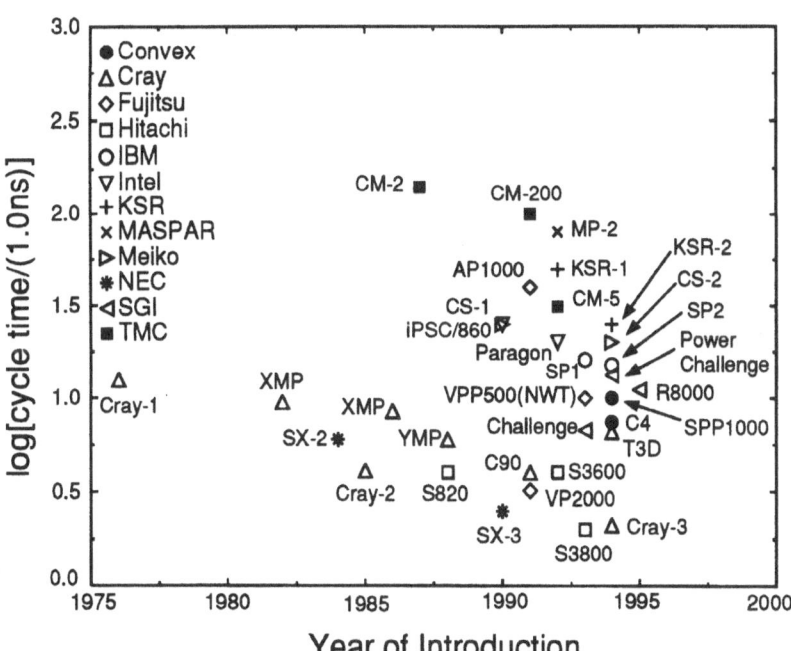

Year of Introduction

year of the Cray 1 delivery) the clock cycle has been reduced from 12.5 ns to 2 ns for the Hitachi S3800, which was first delivered in 1993. There is another interesting trend illustrated in figure 2: it is clear that the traditional custom built vector processors no longer have a substantial lead over off-the-shelf microprocessors. Indeed, many of these mass produced RISC-based processors now operate with clock cycle times of 10 ns or less. For example, the DEC Alpha 21064 microprocessor in the Cray T3D has a 6.6 ns cycle time and the Convex Examplar SPP1000, which utilises Hewlett Packard's Precision Architecture (PA-RISC) processor, has a cycle time of 10 ns. More recently, DEC introduced their new Alpha 21164 microprocessor (not shown in figure 2) which operates at 3.3 ns and the next Cray MPP system, the T3E, will be based on this new chip. For the present, it would therefore appear that the technologies are converging to very similar clock cycle times.

As figure 2 indicates that there have not been any dramatic changes in cycle times, has there been any dramatic improvement in the peak performance? If there has, how has this been achieved? To begin to answer these questions, it is instructive to look at the performance of a single processor. These are shown in figure 3 for a range of representative architectures, including some of the most powerful systems available in 1995. In 1976, the Cray 1 had a peak performance of 160 Mflop/s. However, by the 1990's, the Japanese had single processors with peak performances of 6.5 Gflop/s (NEC SX-3) and 8 Gflop/s (Hitachi S3800), the latter being available in 1993. The Cray 1 had a cycle time of 12.5 ns and the SX-3 and S3800 have 2.5 ns and 2 ns, respectively. Hence, the cycle time has been reduced by a factor of about 6 but the peak performance has increased by a factor of 50. This leads naturally to the second part of the question: how has this been achieved? The answer, of course, is by introducing additional parallelism. The Cray 1 had two vector pipelines whilst the S3800 has eight multifunctional units, each of which can perform a multiply and add operation, and can deliver a staggering 16 results per clock cycle.

Figure 3 also shows that the (RISC-based) microprocessor has only been on the commercial scene in any great quantity since about 1990. However, whilst their peak performance is somewhat less than that of their vector processor rivals, they too have shown dramatic improvements in performance. The IBM SP-2 (first delivered in 1994) has a peak performance of 266 Mflop/s and the Intel iPSC/860 (first delivered in 1990) is theoretically capable of 60 Mflop/s. The DEC Alpha 21164, delivered in 1995, has a peak performance of 600 Mflop/s and other microprocessors, such as the PA-8000 and MIPS R10000, which were announced in 1995, will be capable of even greater performance, probably up to 800 Mflop/s. It is feasible that a 1 Gflop/s (1000 Mflop/s) microprocessor could be available at some point during 1996. In 1985, the fastest microprocessors were capable of about 1 Mflop/s and by 1996 (a little over ten years later) they will have improved by a factor of approximately 1000! This is a phenomenal rate of growth. The other important factor is the price/performance ratio, which continues to decrease. Again, many of the improvements in microprocessor performance have been achieved by the introduction of parallelism at the micro-architecture level. These improvements have generally been accomplished by "borrowing" technology from vector processors e.g. pipelining. Other techniques, such as superscalar and superpiplining, have also been used and these approaches will be explained later. However, it is now argued that the microprocessor industry is beginning to be the innovation leader as superscalar architectures start to use *out-of-order* instruction issue and completion to improve cache management and minimise the effect of data dependencies.

Figure 3 Improvement of single processor peak performance since 1975

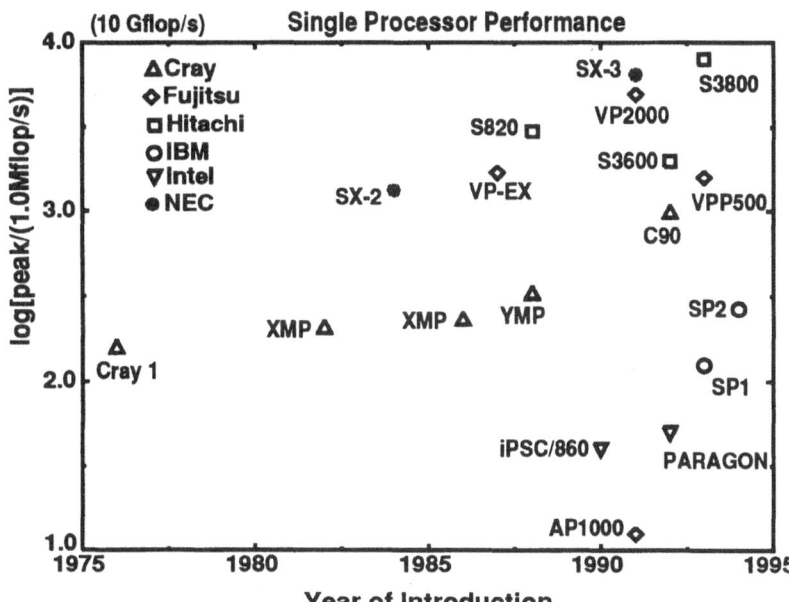

How have these improvements affected the overall performance of a machine? This is illustrated in figure 4 which shows the absolute peak performance of many of today's leading machines. The figure has been compiled from information contained in the TOP500 reports (Dongarra et al) over the last few years. Before proceeding to analyse the figure it is necessary to state that some of the data comes from the vendors specifying the maximum size of system that can be built, such as the Cray T3D which can have up to 2048 processors. As yet, nobody has purchased a fully configured Cray T3D, although systems with 1024 processors have certainly been installed. In some cases there is no limit to the number of processors that can be coupled together and so data for the largest configuration delivered, as reported in the TOP500, has been used. The Intel Paragon and IBM SP2 could be placed in this category. Indeed, Intel have been commissioned to develop a Teraflop machine which will contain over 9,000 of their next generation Pentium Pro (P6) microprocessors and which is scheduled to be installed in late 1996.

Figure 4 clearly shows that the vector supercomputers no longer dominate in terms of peak performance. Many of the top systems belong in the Massively Parallel Processing (MPP) class and contain inexpensive off-the-shelf components. Another factor that is evident is that parallelism is necessary to achieve performances on this scale and the dramatic improvement since 1976, from 160 Mflop/s, to machines bordering on the 1 Tflop/s (1,000,000 Mflop/s) is phenomenal. The fastest machine in the world today (1995) is the Fujitsu VPP500 (also known as the Numerical Wind Tunnel or NWT). The memory is physically distributed and each processor has 256 MBytes of memory. It is, therefore, a DM-MIMD machine with 140 vector processors although it could have up to 222 processors, as indicated by the performance quoted in figure 4. The progress indicated in figure 4 has not been made without cost. In 1994,

two major vendors, Thinking Machines Corporation and Kendall Square Research, filed
for bankruptcy. This was followed in March 1995 by Seymore Cray's company, Cray
Computer Corporation (not to be confused with Cray Research Inc. which is very much
in business). Hence, the supercomputing world is extremely volatile. However, from
the foregoing discussion, it is evident that parallel processing is here to stay and, at
present, remains the only way to achieve high performance in computing.

Figure 4 Peak performance for fully configured systems

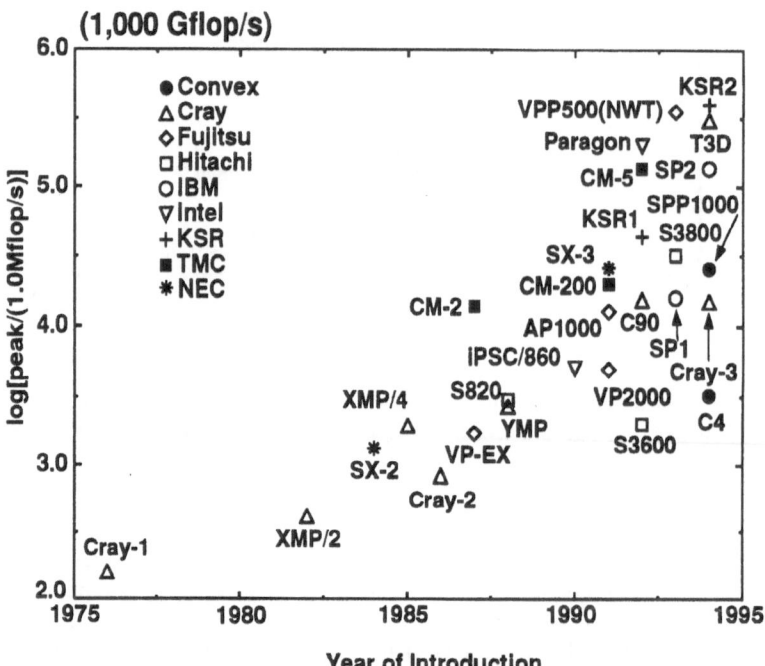

4. Network Topologies

There are many different ways of connecting processors together. The main difference
between the approaches is whether the memory is local or shared. Most architectures
use a shared memory arrangement whereby each processor can access data from a
common memory store e.g. workstations, vector processors like the YMP, C90. This
arrangement is shown in figure 5. Any communication between the processors is via
the global memory and a high speed bus. One problem that can arise with this system
is that more than one processor may wish to access the same memory location. This
is known as *memory contention*. A time delay is then incurred until the memory is
free. Another problem area is *bus contention*, which can become severe with large
numbers of processors.

Figure 5 Typical Shared Memory Interconnect

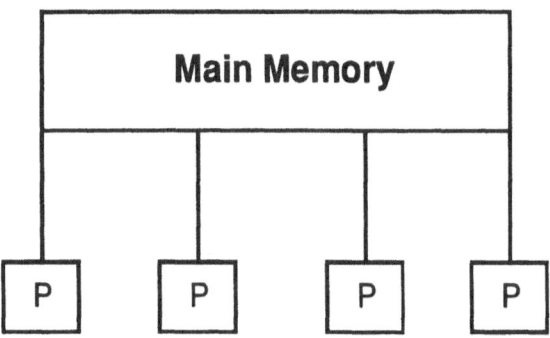

As previously mentioned, distributed memory systems need to communicate with each other. Therefore, any inter-processor communication is done by message passing. Figure 6 shows the interconnect topology for a *completely connected system*. A completely connected system of p processors would each require $p - 1$ connections, which is impractical for large systems. Another method of obtaining a completely connected system is to use a *crossbar switch*, as illustrated in figure 7. Here, each processor has access to all the memory units with fewer connection lines. However, the number of switches required to connect p processors to m memory modules is $p \times m$. Memory contention will also occur if 2 or more processors try to access the same memory.

Figure 6 Completely Connected System

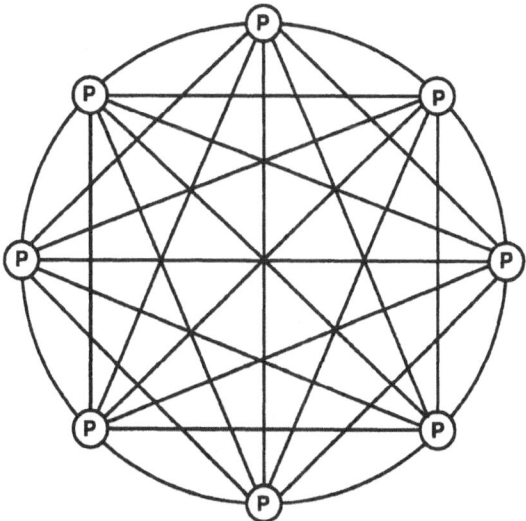

Figure 7 Crossbar Switch Network

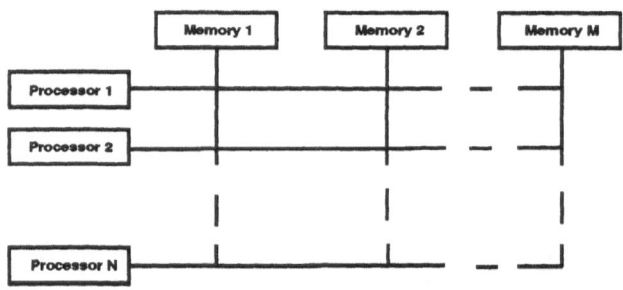

Figure 8 shows a ring network. This also utilises a high speed bus approach. Each message placed on the ring by a processor would contain a destination address and a source address. A popular connection scheme is to employ a mesh arrangement. The simplest example is a *linear array*, as illustrated in figure 8. In this arrangement, each processor is connected to its nearest neighbour (except at the end). Whilst this

Figure 8 The Ring Network and the Linear Array

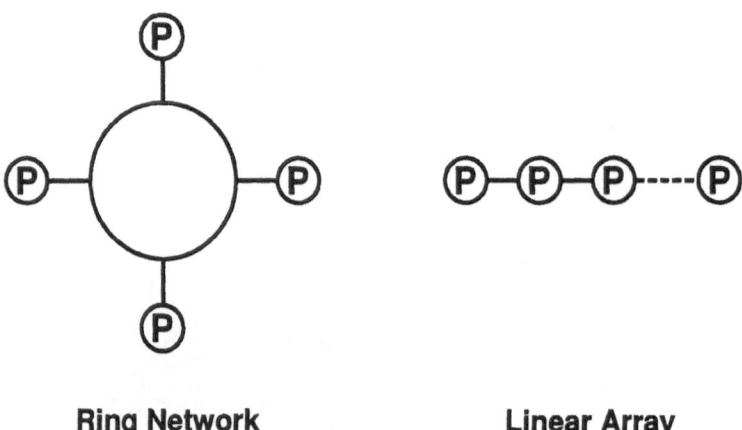

Ring Network   **Linear Array**

has the advantage of simplicity, the disadvantage is that data may need to be passed through $p - 1$ processors. Joining the two end processors to make a simple toroid would enhance its flexibility. The obvious extension of the linear array is a 2–D or 3–D mesh-connected array. The Intel Paragon employs a 2–D mesh and the Cray T3D has a 3–D toroid configuration. Whilst the simplicity is retained, the disadvantage of having to pass messages through intermediate processors remains.

Figure 9 Star and Tree Networks

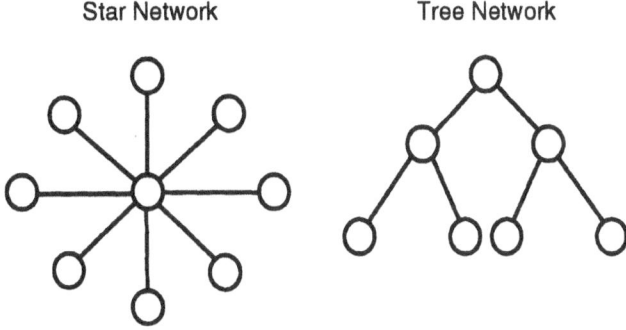

Star Network   Tree Network

One of the more interesting connection topologies is that of the hypercube. The main market vendors for this connection scheme are Intel and nCube. One of the advantages of this approach is that the communication length (which depends strongly upon the routing technology) only grows as $log_2(p)$ as the number of processors increase. The disadvantage is that the complexity and number of lines from each processor also increases and a practical size limit to the cube will be reached. The hypercube arrangement has certainly lost its popularity and the foregoing reason may have been a contributing factor as massively parallel processing became a reality. Other connection networks have been proposed, such as the star and tree networks illustrated in figure 9, but are not used in many practical applications.

At this stage it is worth comparing the various topologies discussed and look at the minimum number of connections needed and what the "worst" case communication would be i.e. what is the maximum number of hops required. To illustrate this, consider the linear array shown in figure 8. For p processors, there are $p-1$ connections needed and the worst case data transfer occurs when the first processor sends a message to the last processor. The maximum number of hops is therefore $p - 1$, also. For a small number of processors this may be okay but for a large number of processors, this would be unacceptable. The tree network illustrated in figure 9 also requires $p-1$ connections but the maximum number of hops only grow as $N_h = 2\{log_2(p + 1) - 1\}$. The completely connected system, as indicated in figure 6, would require $N_c = p(p - 1)/2$ connections but the maximum number of hops for any communication would always be one. For Cartesian mesh topologies, of which the linear array can be considered a 1D mesh, then the number of connections grows as $N_c = 3p - (IJ + IK + JK)$ for a mesh of size (I, J, K) and the number of hops required is $N_h = I + J + K - 3$. For the special cases of square and cubic topologies, the number of connections required is $N_c = 2\sqrt{p}(\sqrt{p} - 1)$ and $N_c = 3p^{2/3}(p^{1/3} - 1)$, respectively, whilst the number of hops is $N_h = 2(\sqrt{p} - 1)$ for the square mesh and $N_h = 3(p^{1/3} - 1)$ for the cubic mesh. It should be noted that the minimum number of connections for a Cartesian mesh is always obtained for the linear array but the least number of hops required for the worst case communication will be for the cubic mesh. Finally, the hypercube requires $N_c = (p/2)log_2p$ connections but the number of hops only grows as $N_h = log_2p$. It is possible to further reduce the maximum number of hops required for Cartesian meshes by developing a torus configuration. For example, the linear torus would only require $\{p - mod(p, 2)\}/2$ hops for a worst case communication. However, the disadvantage is that more connections are needed. It is therefore evident that a compromise has to be made when designing the communication hardware. It should also be emphasised that data transfer is usually between neighbouring processors and, therefore, worst case data transfers are not always important. Another aspect that could be important, particularly if more processors are going to be added, is how the machine grows in size. For example, the hypercube requires that the number of processors is a power of 2 i.e. if a 64 processor system is to be increased, the next available size is 128 processors. The Cray T3D, whilst being a 3D torus, also has this restriction. This could be an undesirable feature if additional processors are required. It should be noted that other connection networks also exist, including omega networks (Lawrie, 1975), indirect binary n-cube arrays (Pease, 1977), and fat trees (Leiserson, 1985).

5. Performance Measurement

This section will be used to determine the relative performance of machines and to discuss aspects such as the *peak* performance. Traditionally, manufacturers would quote a machine's performance in MIPS (Millions of Instructions Per Second). However, the execution of an instruction will vary between different machines and is therefore not necessarily a good indicator of what a machine's floating point performance will be. It has therefore become traditional to quote the machine's peak floating point performance. Unfortunately for parallel processing, very few algorithms have realised a significant fraction of a machine's potential. This has led to the more cynical view that the peak performance of a machine is the speed that the manufacturer guarantees will not be exceeded. Whilst some matrix operations perform well on parallel systems, with regard to obtaining a significant fraction of the peak performance, "real" applications rarely obtain what could be considered a respectable fraction. The Numerical Aerodynamic Simulation (NAS) Parallel Benchmark results (Bailey et al, 1993) show this very clearly with a generic suite of CFD codes which perform virtually no I/O (always a problem on parallel systems) and have removed many of the complexities associated with turbulence and boundary conditions. The typical CFD result here shows that a 128–node Intel iPSC/860 performs on a par with one YMP processor. If the potential of each machine is compared then the iPSC/860 is approximately 16 times that of a single YMP processor.[1] The foregoing calculation for the peak performance of a machine is obtained from:

$$r = \frac{N_r \times N_p}{\tau} \qquad (1)$$

where N_r represents the number of floating point results per clock cycle (usually in 64 bit precision), N_p is the number of processors, and τ is the clock cycle time (usually in nanoseconds). Occasionally the manufacturer gives the clock frequency in MegaHertz (MHz), in which case τ is given by the reciprocal of this value (i.e. $\tau = 1/\text{clock}$ frequency in MHz). In the example given above the Intel has 128 processors ($N_p =$

---

[1]   This assumes that each iPSC/860 processor is capable of 40 Mflop/s in 64 bit arithmetic. Originally, Intel claimed that the processor was capable of 60 Mflop/s which was obtained by pipelining and instruction caching (Trew and Wilson).

128), a clock frequency of 40 MHz ($\tau = 25$ ns), and can deliver one floating point result (an addition, subtraction or multiplication but not a division) every period (N_r = 1, originally N_r was claimed to be 1.5). This gives a peak performance of 5.12 Gflop/s. The YMP has a clock cycle time of 6 ns and each processor can deliver 2 floating point results every period by employing two vector pipelines. The peak performance of each YMP processor is therefore 333 Mflop/s which, as stated earlier, implies the Intel is potentially 16 times more powerful than a single YMP processor. Of course, peak performance calculations ignore start-up times (on pipelining), stripmining effects, and scalar arithmetic on the YMP and, on the Intel, that a perfectly parallel algorithm is used (where messages are transferred and received instantaneously or are unnecessary) and that the cache is always full. If these and other detrimental factors are included, the performance can degrade significantly. The following sections illustrate some of these features.

It is to be noted that there is a considerable difference between the peak performance and "sustained" performance. As previously mentioned, the sustained performance of many parallel systems can be disappointing for many applications. This is particularly true for processors employing RISC (Reduced Instruction Set Computer) technology. There are several reasons why the sustained performance on these computer systems is frequently low and one important factor is having to perform divisions. On the Intel iPSC/860 and Cray T3D, a divide is performed in software and can take a relatively large number of clock cycles to execute. If IEEE compliant results are required, this figure can increase significantly e.g. on the Intel i860, a Newton iteration scheme is used for divisions and requires 38 clock periods but if it is necessary to have the division comply with the IEEE standard, then the cost increases to approximately 190 clock cycles[2] (Bailey, 1993). The cost of executing square roots, logarithms and exponentiation is also very high. Although there are many inner loops that do not require divisions or could have the divisor moved out, there are still many algorithms in popular use that have divisors that are constantly changing and cannot therefore be removed. This is very much the case for Alternating Direction Implicit (ADI) schemes which are used to solve tridiagonal systems of equations. Another important factor in the poor performance of RISC-based processors is the cache memory bandwidth and the cache size. On the Intel, there is an 8 KByte data cache which can transfer two 64–bit operands to the registers at a rate of 80 MW/s. In theory, it can also perform a floating point operation on data already stored in the registers which, with a clock rate of 40 MHz, gives a theoretical peak performance of 60 Mflop/s in 64–bit arithmetic (in practice, this doesn't happen and it is more typical to quote 40 Mflop/s as the peak performance). However, the data transfer rate from main memory is only 16 MW/s and this results in some disappointingly low performance figures on the Intel i860. The DEC Alpha 21064 processor is currently employed by Cray on their new massively parallel system, the Cray T3D. It is IEEE compliant and it also features an 8 KByte data cache and an 8 KByte instruction cache. The cache bandwidth is 150 MW/s, almost twice the i860 bandwidth.

5. 1. PERFORMANCE CHARACTERISTICS

Before analysing the performance of parallel systems, it is instructive to review

---

[2]   In practice, many people use the —Knoieee flag as a compile option on the Intel and Cray vector systems do not perform IEEE arithmetic.

some well known features of vector processing. From equation 1 it follows that when N million floating point operations are carried out in t seconds, the computational speed is:

$$r = \frac{N}{t}(Mflop/s)$$ (2)

and it follows that the computational time required is therefore:

$$t = \frac{N}{r}(microseconds)$$ (3)

However, in most practical situations not all operations can be executed with high computational speed i.e. there will be a scalar and vector part to the operation. If we make the assumption that the algorithm operates at r Mflop/s and that some fraction (f) of this work is carried out in an "idealised" vector operation at V Mflop/s and the rest of the computation is carried out in scalar mode at S Mflop/s, then the time required for the computation is given by:

$$t = t_V + t_S = \frac{fN}{V} + \frac{(1-f)N}{S} = N\left(\frac{f}{V} + \frac{1-f}{S}\right)$$ (4)

The overall performance is therefore given by:

$$r = \frac{N}{t} = \frac{1}{f/V + (1-f)/S}(Mflop/s)$$ (5)

Equation (5) is known as Amdahl's Law (Amdahl, 1967) and some important implications can be drawn from this. If $V \gg S$ then the time for the vector operation is small and can be ignored. The time for the computation is therefore:

$$t \geq \frac{(1-f)N}{S}$$ (6)

If the algorithm had operated at the lower speed the computational time would have been $t = N/S$ (microseconds). Hence, the relative gain in time from executing part of the algorithm at speed V, instead of at the much lower speed S, is:

$$G = \frac{t_1}{t_2} = \frac{N/S}{(1-f)N/S} = \frac{1}{1-f}$$ (7)

and the relative gain is therefore bounded by equation (7). This means that f has to be quite large to obtain a significant improvement in performance and this is illustrated in figure 10 (n.b. on the AMDAHL VP 1100 and VP 1200 series, V/S was 28.5 and 57, respectively).

Figure 10 Amdahl's Law For A Range of V/S Ratios

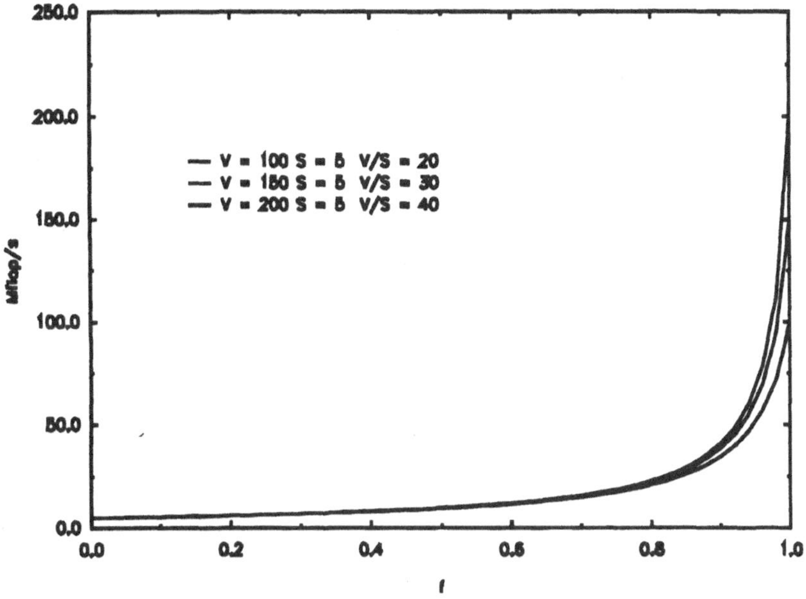

6. Speed-up, Efficiency and Scalability

As previously mentioned, many people are interested in reducing the wall clock time of a particular computation. This brings in the concept of *speed-up*. If the computations were carried out in p equal parallel parts then the wall clock time would be reduced ideally by a factor of $1/p$. If the wall clock time is denoted by t_p for a given job to execute on p processors, then the speed-up is defined as:

$$S_p = \frac{t_1}{t_p} \qquad (8)$$

The speed-up is a measure of how the algorithm compares with itself on 1 and p processors. If a job took 60 minutes on a single processor and 15 minutes on 4 processors, then the speed-up would be 4. This is the ideal case. Under certain circumstances, it is also possible to obtain a *super linear* speed-up. This usually occurs when the sparsity of a matrix is exploited on a parallel architecture or advantageous caching occurs (Emerson et al, 1993). It should be noted that there is currently some debate as to which sequential time the parallel computation should be compared to. For example, a highly vectorised code would have long vector lengths which is ideal for a YMP etc. but, in general, it is not suitable for a cache based architecture. Moreover, parallel codes will have additional logic to identify which part of the domain they are operating on. Hence, a comparison of the parallel code's performance on a single processor and on multiple processors is one measure of the speed-up attained but, for a given problem size, it may still be slower than the vectorised code. Many of the issues involved in reporting results have been discussed (see Bailey 1991, 1992) and appropriate care should be taken when such information is reported.

Another important factor is the *efficiency* of the parallel implementation. This is defined as:

$$E_p = \frac{S_p}{p} \qquad (9)$$

For the case where the algorithm has a perfect speed-up, the efficiency is 1. In general, the efficiency lies in the range $0 \leq E_p \leq 1$ although efficiencies greater than one will arise if a super linear speed-up is obtained. The speed-up and efficiency are obviously closely related and give and indication of how well balanced the computational load is and how the problem scales. In practice, a perfect speed-up is rarely attained and there are several reasons for this, which are:

(i) The algorithm may not be perfectly parallel i.e. certain algorithms may have an inherently serial part;
(ii) The load balancing of the processors may not be perfect. When a code is run in parallel, each processor is assigned an amount of work to do. If one processor is required to do more work than another processor it will take longer to complete its task;
(iii) There will be a timing cost for any communication between processors. In general, when a program is run in parallel, the processors will have to send and receive messages from other processors. Each time a message is sent, an overhead in timing cost is incurred;
(iv) The parallel code may need to be synchronised after certain operations.

Amdahl's law can also be applied to parallel computing. As with vector computing, in most practical situations not all operations can be executed in parallel i.e. there will be a serial part to the algorithm. Following the approach established for vector processing it is assumed that all operations are now carried out at the same computational speed and that a percentage (f) is carried out in parallel on p processors whilst the rest of the work $(1 - f)$ has to be carried out on one processor, then the wall clock time will be:

$$t_p = t_{\shortparallel} + t_s = \frac{ft_1}{p} + (1 - f)t_1 \geq (1 - f)t_1 \qquad (10)$$

Hence, the speed-up is given by:

$$S_p = \frac{t_1}{t_p} = \frac{p}{f + (1 - f)p} \qquad (11)$$

It can be seen that the speed-up is reduced by a factor $f + (1 - f)p$. This relationship is sometimes known as Ware's law (Ware, 1973). From this expression for speed-up, the efficiency is given by:

$$E_p = \frac{S_p}{p} = \frac{1}{f + (1 - f)p} \qquad (12)$$

Similar implications from Amdahl's law for parallel processing can again be drawn. If a very large number of processors were available, then the speed-up would be bounded by:

$$S_p \leq \frac{1}{1 - f} \qquad (13)$$

and the speed-up is limited by the proportion of work carried out in serial. It is therefore necessary to get f close to 1 to optimise the performance. Even with f close to 1 there are significant reductions in speed-up for increasing values of p. Table 2 illustrates the performance deterioration for various values of p and f.

Table 2 Parallel performance implications of Amdahl's law

f	p=1	2	4	8	16	32	64	∞
1.00	1.00	2.00	4.00	8.00	16.00	32.00	64.00	∞
0.99	1.00	1.98	3.88	7.48	13.91	24.43	39.26	100.00
0.98	1.00	1.96	3.77	7.02	12.31	19.75	28.32	50.00
0.96	1.00	1.92	3.57	6.25	10.00	14.29	18.18	25.00
0.94	1.00	1.89	3.39	5.63	8.42	11.19	13.39	16.67
0.92	1.00	1.85	3.23	5.13	7.27	9.19	10.60	12.50
0.90	1.00	1.82	3.08	4.71	6.40	7.80	8.77	10.00
0.75	1.00	1.60	2.28	2.91	3.37	3.66	3.82	4.00
0.5	1.00	1.33	1.60	1.78	1.88	1.94	1.97	2.00

One objection to the very pessimistic projection from Amdahl's law is that it implicitly assumes that the proportion of parallelised (or vectorised) code (given by f) is independent of the problem size or vector length. If you were to take a fixed size problem and run it on an increasing number of processors, then the speed-up bound would come into effect. However, when using a more powerful processing system, it is generally observed that as the problem size is scaled with the number of processors, the efficiency remains approximately constant. Gustafson (1988) proposed that it might be more realistic to consider the *run time* and not the problem size. He assumed that the problem could be solved in one unit of time on a parallel machine ($t_p = 1$) and that the time for the serial computation was $(1 - f)$ and for the parallel computation was f. With these assumptions, if there are p processors available then the time for a single processor to execute would be:

$$t_1 = (1 - f) + pf \tag{14}$$

Hence the speed-up is given by:

$$S_p = \frac{t_1}{t_p} = p - (p - 1)(1 - f) \tag{15}$$

In the work performed by Gustafson on a 1024 processor Ncube system, where the sequential part of the code varied from $0.4 - 0.8$ percent for a single node problem (which included the overhead from program startup, message passing, I/O), the speed-up varied from 1016 (for a flux corrected transport code) to 1021 (for a conjugate gradient solver). Amdahl's law would suggest that the maximum possible speed-up that could be attained is 250! Has Amdahl's law been violated? In reality, no, but it does indicate that the predictive capabilities of Amdahl's law are very limited. The interpretation of Amdahl's law by Gustafson has been challenged by Heath and Worley (1989) who recognised that the sequential part of the computation quoted by Gustafson reduced as the problem size increased and that had the serial part for the larger computation been used, Amdahl's law would give the correct speed-up. However, this misses the point that Gustafson was trying to make concerning the lack of predictive value from Amdahl's law.

The subject of scalability is becoming increasingly important with the advent of massively parallel processing (MPP). As hundreds, or even thousands, of processors can now be linked together, how a problem performs when it is scaled up to utilise these large numbers of nodes is critical. It may be that many current algorithms may need to be reconsidered if MPP systems are to be used effectively and efficiently. Indeed, it is already considered necessary to distinguish between architectural scalability and algorithmic scalability (Nussbaum and Agarwal, 1991). In many practical applications, particularly in implicit CFD algorithms, the number of iterations required to obtain a converged solution of a sequential code generally increases as the problem size is increased. This problem can become worse in parallel because the partitioning can also adversely affect the number of iterations required when compared to the sequential case. In time accurate simulations, the CFL condition is related to the grid size and the allowable time step will therefore be affected when the grid size is increased. This will therefore necessitate more time steps to reach some arbitrary point in time.

Many vendors currently claim to have scalable architectures but this is a claim that is very difficult to prove or disprove. This is primarily because there is no formal definition of scalability (Hill, 1990). The development of scaling models has been discussed by Singh et al (1993) whereby three approaches are reviewed, namely: constant problem size (limited by Amdahl's law); memory-constrained scaling (as followed by Gustafson); and time-constrained scaling. An important feature identified by Singh et al was how the communication-to-computation ratio behaves as the problem is scaled. It is clear that a coherent scaling methodology needs to be developed, particularly for real applications, that can accurately evaluate MPP systems.

7. Pipelined, Superpipelined and Superscalar Architectures

There are now many ways of trying to get extra performance from processors. One of the oldest is pipelining: this has been the main approach used by vector processors to get improved performance and pipelining has also been embraced by the microprocessor designers. The high performance is obtained by overlapping the instructions, as illustrated schematically in figure 11 which shows a simple 4–stage pipeline. In reality, the number of stages is higher than that indicted in figure 11 e.g. the DEC Alpha 21164 has a 9–stage floating point pipeline (Edmondson et al). The objective of pipelining is to deliver one result every clock cycle and this is achieved once the pipeline is full, although there is some overhead incurred in filling up the pipeline. Most pipelines are static i.e. they perform a fixed task, such as multiplication, although dynamic or reconfigurable pipelines are possible. In general, most architectures employ multiple pipelines and it is common for multiplication and addition pipelines to be present. Vector processors frequently employ special pipelines that are capable of performing division or square root operations. These operations are very expensive in RISC-based systems and may be performed in software. However, these operations are now beginning to be performed in hardware as well. Examples of architectures that use pipelining to improve performance are the NEC SX-3 vector processor and the DEC Alpha 21064 microprocessor.

Figure 11 Instruction scheduling in a pipelined processor

Pipeline cycle
F
D
E
W
F = Fetch, D = Decode, E = Execute, W = Write

Figure 12 Instruction scheduling in a superpipelined processor

Master clock cycle
Pipeline cycle
F
D
E
W
F = Fetch, D = Decode, E = Execute, W = Write

A pipelined processor can be divided up into a number of substages and this is illustrated in figure 12. This is known as *superpipelining*. The number of substages involved represents the *degree* of superpipelining and two substages are shown in figure 12. The substages have a faster clock cycle as indicated in the figure, where the pipeline cycle time is half of the master cycle time. This effectively means that when the pipeline is full, it is possible to obtain one and a half results every clock cycle.

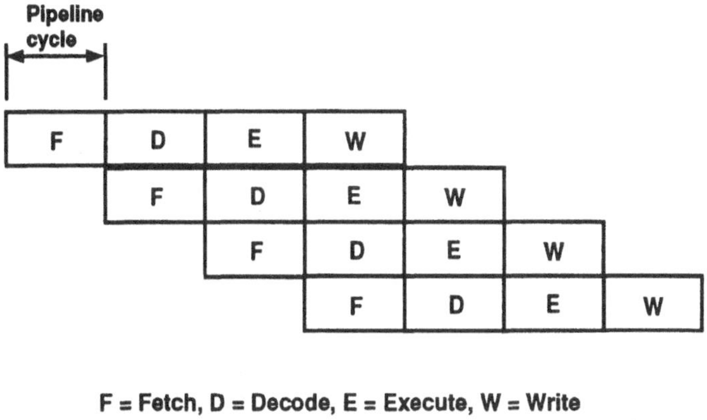

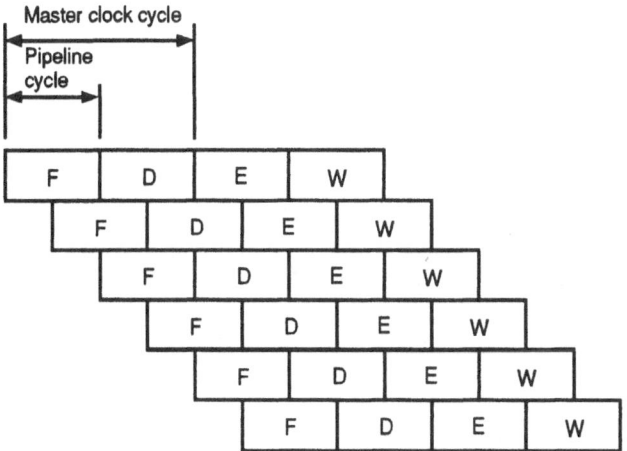

Hence, for every 2 master clock cycles that elapse, three results are available. This technology has been employed in the MIPS R4000 processor (Mirapuri et al, 1992). The disadvantage with this approach is that some overhead is incurred in having to monitor the substages. The more substages involved, the higher the overhead.

Figure 13 Instruction scheduling in a superscalar processor

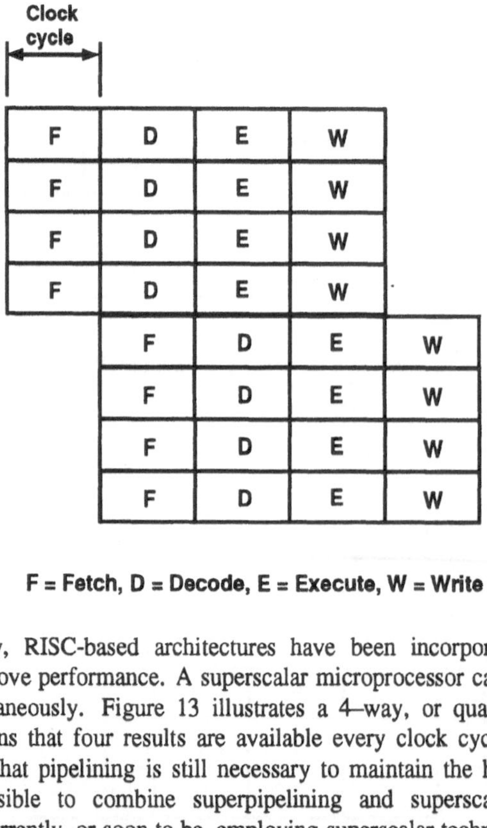

F = Fetch, D = Decode, E = Execute, W = Write

More recently, RISC-based architectures have been incorporating superscalar technology to improve performance. A superscalar microprocessor can execute several instructions simultaneously. Figure 13 illustrates a 4–way, or quad-issue, design.In principle, this means that four results are available every clock cycle. An important feature to note is that pipelining is still necessary to maintain the high performance. Indeed, it is possible to combine superpipelining and superscalar technologies. Microprocessors currently, or soon to be, employing superscalar techniques are Hewlett Packard's PA-8000, IBM's RS/6000 and DEC's Alpha 21164. All of the foregoing execute up to four instructions every clock cycle.

8. Memory and Cache Organisation

Data locality plays an important role in high performance computing. This is particularly true when a cache-based system is being used because a cache miss is very expensive in terms of lost cycles. This is because when your program makes a data reference, it will first look in cache to see if the data requested is there. If it is, it can be obtained very quickly. If it is not, a cache miss occurs and it may be necessary to go to main memory to fetch the requested data. The bandwidth between main memory and cache is usually three to five times less (or even worse) than the cache bandwidth. For example, the DEC Alpha cache bandwidth is 150 MW/s but its main memory bandwidth is only 37 MW/s (Bailey, 1993). Also, it is worth noting that when a cache miss occurs, a *cache line* is replaced. This usually consists of 4 or 8 integers or real numbers and one of these numbers will be the requested data item and the others will be from neighbouring memory locations. This is because the cache will assume that these are the most likely future requests. These problems arise because the cache size is smaller than that of main memory. Locations in main memory then have to be *mapped* into the cache. In order to minimise cache problems it is beneficial to have an understanding of the most commonly used cache mappings, which are: (i) direct mapped; (ii) set associative; and, (iii) fully associative. The foregoing grouping also reflects the increasing complexity and cost of cache memory. Many contemporary systems use one of these approaches, for example, the Cray T3D employs a direct mapped cache, the IBM SP2 wide node has a 4–way set associative cache and the MIPS R10000 has a 32 KByte 2–way set associative on-chip primary cache with a 2–way set associative off-chip secondary cache ranging in size from 512 KBytes to 16 MBytes. A secondary cache helps to minimise the effect of a primary cache miss and prevent having to access main memory which, as described before, is a relatively slow process. It is also possible to have additional memory hierarchies, such as a tertiary cache. The cache line that is replaced is generally determined by the *least recently used* (LRU) replacement algorithm.

8. 1. DIRECT MAPPED CACHES

This is the simplest of all cache organisations. In this system, a location in main memory has a direct mapping with the cache memory. This is illustrated in figure 14 and the diagram indicates that a conflict can occur when cache line *i* requires data from more than one of the locations in main memory that map directly to line *i*. Let us

Figure 14 A 256 line direct mapped cache and memory layout

A	B

| line 0 |
| line 1 |
| |
| |
| |
| |
| |
| |
| |
| |
| |
| |
| |
| line 255 |

Cache Memory

A	B
maps to line 0	maps to line 0
maps to line 1	maps to line 1
maps to line 255	maps to line 255

Main Memory

illustrate this further with a practical example: the Cray T3D has an 8 KByte direct mapped data cache arranged as 256 lines, with each line containing 4 Words (32 Bytes). The following code fragment would therefore cause cache *thrashing* and would result in very poor performance because every time that it tried to load an element from A, say A(1), it would also need the corresponding element from B, that is B(1), and a cache miss would occur. The cache would then need to reload data for B. This would occur for **every** element of both the arrays.

```
REAL A(1024),B(1024)

COMMON /ARRAY/A,B

DO I = 1 , 1024
   A(I) = A(I) + B(I)
ENDDO
END
```

Fortunately, a simple remedy exists to prevent this occurring. By "padding" the arrays with a dummy array of appropriate size, the mapping can be made to avoid thrashing. In the foregoing example, an array of 4 elements would be sufficient i.e.

```
REAL A(1024),B(1024),DUMMY(4)
COMMON /ARRAY/A,DUMMY,B
```

In this case, the elements B(1) - B(4) would then map to cache line 1 and any conflict would be avoided. A better solution still would be to set the dummy array to DUMMY(512) because, whilst the array elements A(1 – 4) have no conflict, A(5) will map to cache line 1 and it will be necessary to empty that cache line of the B elements. This process will then be repeated throughout the loop. However, by making the array size equal to 512 elements, the cache will be able to load up 128 lines of arrays A and B. Of course, this is a simple example to illustrate a point and in practice, it is very difficult to spot this problem by the use of most profiling tools. It is more likely that the programmer will realise that something is wrong because of the long execution time.

8. 2. FULLY ASSOCIATIVE CACHES

This is the best but most expensive cache organisation. It allows any location in main memory to map into any location in cache memory. However, the disadvantage to this cache is that it takes longer to search for a particular data item in main memory because it can reside anywhere rather than in specific blocks. Also, due to their complexity, they are generally smaller.

8. 3. SET ASSOCIATIVE CACHES

A set associative cache is essentially a compromise between a direct mapped cache and a fully associative cache. It is generally designed to be 2–way or 4–way set associative and, as illustrated in figure 15, a 2–way model will be used to explain how it works. A simple way to visualise how a 2–way set associative cache operates is to consider two direct mapped caches side by side. The MIPS R10000 32 KByte cache, an example cited previously, is actually arranged as two 16 KByte caches each containing 256 lines, with each line holding 8 Words (64 Bytes). To indicate the same problem, however, it is necessary to increase the array size to 4096 elements i.e. A and B are now 32 KBytes. In the code fragment used earlier, elements A(1) and B(1) mapped to the same cache line and caused cache thrashing. In this approach, element A(1) can map to either line 0 in cache 1 or line 256 in cache 2 and similarly for B(1). Hence, the thrashing illustrated previously is avoided. However, as stated, a set

Figure 15 A 2–way set associative cache and memory layout

Cache 1	Cache 2		A	B
line 0	line 256		maps to line 0 or 256	maps to line 0 or 256
line 1	line 257		maps to line 1 or 257	maps to line 1 or 257
line 255	line 511		maps to line 255 or 511	maps to line 255 or 511

Cache Memory   Main Memory

associative cache is a compromise and thrashing would occur in the following loop because C(I) will have nowhere to map to without causing a conflict. Again, suitable padding will avoid this problem.

```
REAL A(4096),B(4096),C(4096)

COMMON /ARRAY/A,B,C

DO I = 1 , 4096
   A(I) = A(I) + B(I) + C(I)
ENDDO
END
```

It is evident that a 4–way set associative cache will be less prone to the simple cache conflicts identified.

9. Optimisation for RISC Systems

Many vendors are now using "off-the-shelf" components based on the Reduced Instruction Set Computer (RISC) methodology. The modern processors are, in general, theoretically capable of performing at least one floating point operation every clock cycle. Some perform multiple floating point operations by employing a Complex RISC (CRISC) approach which allows for more involved instructions to be used, such as an addition and a multiplication, or by employing superscalar or superpipelined operations. Where RISC based systems begin to fall down is on other types of arithmetic operations, such as division, exponentiation, etc. It is therefore worthwhile remembering that these operations should be avoided or at least kept to a minimum, particularly if they occur in a numerically intensive part of the code. As a simple example, consider the decoding of a column vector $Q = (\rho, \rho u, \rho v, \rho w, E)^T$ to find the local velocity components u, v, and w. This occurs frequently in computational fluid dynamics. One way of doing this would be the following:

$$U = Q_2/Q_1$$
$$V = Q_3/Q_1 \tag{16}$$
$$W = Q_4/Q_1$$

but it would be better accomplished as follows:

$$R = 1.0/Q_1$$
$$U = R * Q_2$$
$$V = R * Q_3 \tag{17}$$
$$W = R * Q_4$$

As previously stated, a divide is generally performed in software and can take a relatively large number of clock cycles to execute. If IEEE compliant results are required, this figure can increase significantly. It can therefore be advantageous to disable the requirement for an IEEE operation at compile time.

A critical factor on RISC systems is memory access. In the ideal situation, you would want all of your data to reside in cache. This, of course, is impossible. It is therefore necessary to minimise the number of cache misses that occur. The best way of achieving this is to have a unit stride between successive elements. Consider the following loop:

```
DO I = 1 , N
   DO J = 1 , N
      A(I,J) = A(I,J) + B(I,J)
   ENDDO
ENDDO
```

This loop has a stride of N and its performance will be relatively poor and will deteriorate as N gets larger. However, the following loop:

```
DO J = 1 , N
   DO I = 1 , N
      A(I,J) = A(I,J) + B(I,J)
   ENDDO
ENDDO
```

will perform well because it has a unit stride. This is because when your program makes a data reference, it will first look in cache to see if the data requested is there.

If it is, it can be obtained very quickly. If it is not, a cache miss occurs and it is then necessary to go to main memory (or secondary cache if such a memory hierarchy is being used) to fetch the requested data. The bandwidth between main memory and cache is usually three to five times less (or even worse) than the cache bandwidth. For example, the DEC Alpha cache bandwidth is 150 MW/s but its memory bandwidth is 37 MW/s (Bailey, 1993). A cache miss can therefore be very expensive. Also, it is worth noting that when a cache miss occurs, a cache line is replaced. This usually consists of 4 or 8 integers or real numbers and one of these numbers will be the requested data item and the others will be from neighbouring memory locations. This is because the cache will assume that these are the most likely future requests. The rest of the cache line will only be used if the stride is sufficiently small to allow that data to be accessed and this is clearly best done if the stride is unity.

Another factor that can help in improving the performance is loop unrolling. A loop for a finite difference stencil can have a large stride when it involves derivatives in several dimensions. It was explained in the preceding paragraph that a cache miss meant that a new cache line was brought from main memory with neighbouring memory locations. Loop unrolling can therefore assist by using neighbouring data items. A typical 3D array in CFD could be stored as Q(IMAX,JMAX,KMAX,5) where Q could represent the column vector previously described. This type of array structure would work very well on a vector processor (providing the stride was okay). However, on RISC architectures it would not be the optimal way of storing the array. A typical vector loop would be:

```
DO L = 1 , 5
   DO K = 1 , KMAX
      DO J = 1 , JMAX
         DO I = 1 , IMAX
            Q(I,J,K,L) = Q(I,J,K,L) + . . .
         ENDDO
      ENDDO
   ENDDO
ENDDO
```

A better approach for RISC systems would be to store what is currently the outer element in the first column i.e.

```
DO K = 1 , KMAX
   DO J = 1 , JMAX
      DO I = 1 , IMAX
         DO L = 1 , 5
            Q(L,I,J,K) = Q(L,I,J,K) + . . .
         ENDDO
      ENDDO
   ENDDO
ENDDO
```

This will allow more neighbouring elements to reside in cache. To enable the cache to fully exploit the data locality, loop unrolling can be performed. This will expose the computations to the compiler. Loop unrolling can be performed on any part of the loop. The primary concern with loop unrolling is providing better memory access. For inner loop unrolling the resulting code would look like the following:

```
DO K = 1 , KMAX
   DO J = 1 , JMAX
      DO I = 1 , IMAX
         Q(1,I,J,K) = Q(1,I,J,K) + . . .
         Q(2,I,J,K) = Q(2,I,J,K) + . . .
         Q(3,I,J,K) = Q(3,I,J,K) + . . .
         Q(4,I,J,K) = Q(4,I,J,K) + . . .
         Q(5,I,J,K) = Q(5,I,J,K) + . . .
      ENDDO
   ENDDO
ENDDO
```

The best performance is usually obtained when the loop is unrolled to the depth of the cache line. However, each loop should be considered and thoroughly tested to see what improvements can be obtained. A disadvantage with loop unrolling is that the code can become more cumbersome and, in some cases, more difficult to read and maintain.

Of course, an intelligent compiler will spot many of these features and try to optimise the code by loop unrolling and loop restructuring. However, this puts a large onus on the compiler and in reality, it the programmer who knows his own code best. It is therefore up to the programmer to assist the compiler wherever possible to get the best performance from a particular code. However, as with all of these approaches, always check to make sure that what you have done is actually beneficial. Sometimes the compiler knows how to do it much better.

10. Parallel Programming Models

The two basic models used in parallel processing are the Single Program Multiple Data (SPMD) model and the Multiple Program Multiple Data (MPMD) model. The most commonly used model in scientific programming is the SPMD model. In this approach, each processor runs an **identical** program but only solves its own set of data. To solve the global problem, the computational domain needs to be distributed over all the processors and this is usually done via domain decomposition. This will be dealt with more fully in another section. The problem really is for each processor to know which part of the domain it is operating on. However, each processor does have access to information about itself and this can be used, for example, to identify who is a neighbour. This is necessary to be able to send data between nodes. A simple example of a SPMD program is the following:

```
IAM = MYNODE()
IF (IAM.EQ.0) X = 1.0
IF (IAM.GT.0) X = REAL(IAM)
```

In this fragment, MYNODE() is an intrinsic Intel routine which returns a number from 0 to $p-1$, where p is the number of processors. Other systems have their own intrinsic function to return this value e.g. on the Cray T3D it is MY_PE(). The values of X will therefore be 1.0, 1.0, 2.0, 3.0 for a four processor machine. The important point is that each node sees the exact same code (a single program) but can operate on different sections of the code (multiple data). As the MPMD model is not that applicable to CFD it will not be discussed any further.

11. Message Passing Models

One of the most critical aspects of parallel processing is the method of communication. In SM-MIMD systems, where the processors are tightly coupled, any data transfer is done by a high speed bus. However, in DM-MIMD systems, message passing is done explicitly. In this case, there are several important factors which play a decisive role in any program performing efficiently: notably the latency (the time required to set up the message), the bandwidth of the interconnect (the data transfer rate), and the connectivity of the system (how many neighbours does each node have). There are two modes of operation for communicating: synchronous and asynchronous. A message broadcast would operate in synchronous mode. This is the normal mode of operation in SIMD array processing where each processor operates in lock-step mode via the control unit. In DM-MIMD machines, the synchronous mode would be used for global communications i.e. broadcasting a value to all processors. Asynchronous communication is a useful facility to use when requests can be issued dynamically. This can allow *latency hiding* by allowing computations to continue after issuing a send and receiving the data when it becomes available, instead of waiting in idle mode. In asynchronous mode, the program makes use of the most recent data and can significantly reduce the communications cost. These two methods can clearly be combined and various aspects of message handling will now be discussed in more detail.

The explanation is easier to understand by an analogy. Consider 2 office workers in different countries. If office worker A wants to get information quickly to office worker B then he could send a fax. To be able to send the fax, office worker A needs to provide a number to identify the destination (in this case the telephone number), the name of the person who is to receive the information, the name and address of who sent the information, and how long the message will be (the number of pages). To send messages between processors on a distributed memory system you also need to identify the message, include the message to be sent, where it is going and who sent it, and specify how long the message will be. The routine that provides this functionality on the Intel is called CSEND. The routine would be called as follows:

```
CALL CSEND ( msg_id , message , msg_len, your_id , my_id )
```

where msg_id is a unique tag (typically an integer number) that identifies the message, message is the information to be sent, msg_len is the length of the message (on the Intel, this is the number of Bytes), your_id is an integer that specifies the destination, and my_id is an integer that identifies the sender (for the RX node on the Intel, this is always set to zero). As previously mentioned, there are different types of message handling available and these will now be described.

Blocked Message Handling: when a process halts further execution of instructions until a message is sent or received, it is called blocked message handling. This is the most common type of message passing. When a process issues a request to receive a message the process executes no further instructions until a message of the correct type has been received. If the message has not yet arrived at the issuing node, the process will remain idle until the message arrives. If the message was already at the node before the process issued the request, the process would only halt whilst the message is put into memory. Similarly, if the process wanted to send a message, no further instructions would be processed until the message was copied from memory into the message-passing network. This does not mean that the message has reached its destination, only that it is on its way. This send operation is *locally blocking* and the memory that stored the message can now be used again. A *fully synchronous* approach would wait until the receiving node sent a message to indicate that the data had arrived. Many applications require the use of synchronous message passing and it is also useful for the initial development of parallel codes because it provides a loose form of synchronisation between processes.

Non-Blocked Message Handling: If the office worker analogy is reconsidered, then a blocked message would correspond to office worker B stopping work until the fax arrives. A more realistic situation is when office worker B continues working on other projects and starts working on the expected document when it arrives. This corresponds to *non-blocked* (or asynchronous) message handling. It is usually more efficient to handle messages asynchronously. An asynchronous receive would allow processes to alert the operating system that messages are expected and they should be delivered and processed as soon as they arrive. In this approach the processor is not idle until the message arrives but continues to execute instructions until the information in the message is required. If, at that point, the message has not arrived, the process will wait until the message arrives. A process can also send messages asynchronously. The process notifies the operating system that it wants to send a message but the process does not wait until the message has been sent. It can then continue to process instructions following the send command. Another method of handling messages is called *interrupt* message passing. This is similar to asynchronous message handling whereby the process continues to execute instructions while waiting for the message to arrive. However, the method differs by interrupting the process as soon as the message arrives so that it can be processed immediately. The process can then return to what it was doing when the interruption occurred. This type of message passing is only available on some MIMD machines.

12.   Communicating on Distributed Memory Architectures

Message passing on parallel platforms is done on a point-to-point basis and the different ways of sending messages has been dealt with in the previous section. However, there are certain aspects that need to be explained to avoid a problem known as *deadlock*. This can occur in several ways and, unfortunately, it can vary from machine to machine i.e. unless a systematic approach is followed, a code that works on machine A will not necessarily work on machine B - and it is not always obvious why! The fault generally lies with the programmer making an assumption, sometimes unknowingly. A typical error in this case is making a demand on the available buffer space for a message to be placed. This could result in a program working for a small problem size but failing when the problem size is increased. A few examples will now be presented using the standard calls for the Intel iPSC/860 although the translation to PVM, MPI or other native message passing library is straightforward. Before proceding, however, it is necessary to set up a few preliminary values. In any parallel program, this will need to be done before message passing can take place. For the example to come, the following information is required:

```
IAM = MYNODE()
NPROC = NUMNODES()
IDS = IAM + 1
IF (IAM.EQ.NPROC-1) IDS = 0
ITAGS = IAM + 1001
ITAGR = ITAGS - 1
IF (IAM.EQ.0) ITAGR = ITAGS + NPROC - 1
X = 4.0 + REAL(IAM)
Y = 0.0
NBYTE = 4
```

This code fragment uses some intrinsic functions, namely MYNODE() and NUMNODES(), which returns a unique node number and the total number of nodes being used to solve the problem. It is then necessary to do some book keeping. For the sample problem we will assume four nodes are available i.e. NPROC = 4 and IAM will takes the values 0, 1, 2, or 3. The next step is to identify which nodes are going to have data sent to them. For this example, we shall assume only near neighbours will talk to each other except at the boundary, where we can use a periodic condition i.e. node 3 will exchange data with node 0. The flag IDS will perform this task and will have the values 1, 2, 3, 0 as the destination nodes for the message. It is now necessary to set up a unique number to send the data to allow the receiving processor to know that it has the correct message. This is done by defining a send flag (ITAGS) and a corresponding receive flag (ITAGR), corrected for the boundary node. The data to be sent is then created (e.g. X) and the message size in Bytes, which for this example will be 4.

The program is now in a position to begin sending and receiving data. On the Intel, this is done with the CSEND and CRECV routines and the data to be received is copied into Y. In all of the following examples, the set up just described will be assumed. The first case that will be considered is when **all** processors issue a receive simultaneously i.e.

```
CALL CRECV(ITAGR,Y,NBYTE)
CALL CSEND(ITAGS,X,NBYTE,IDS,0)
```

This case will fail because a blocking receive has been issued before any data has been sent. It will therefore cause the program to hang. What happens if we reverse the order of sending and receiving? i.e.

```
CALL CSEND(ITAGS,X,NBYTE,IDS,0)
CALL CRECV(ITAGR,Y,NBYTE)
```

In this case, all processors attempt to send a message at the same time. On the Intel, this will work. However, it is potentially unsafe because for large messages it could swamp any available buffer space. Could the first approach be used with an asynchronous receive? In this case, we need to "post" a receive and when the data becomes available, we can collect it i.e.

```
MSGID = IRECV(ITAGR,Y,NBYTE)
CALL CSEND(ITAGS,X,NBYTE,IDS,0)
CALL MSGWAIT(MSGID)
```

In this approach, the processor states that it wants to receive the appropriate data when it becomes available but until then, carry on doing something useful. For this example, the processors begin to send their data but they could equally have done some other calculation. Again, this example is potentially unsafe because of the requirements being placed on the buffer space. This example works on the Intel and could work on many systems, particularly if the send operation were to occur at a different time on each processor because the program was not synchronised. The safest way to transfer data, however, is to alternate tasks between nodes. This can be done in the following way:

```
IF (MOD(IAM,2).EQ.0) THEN
   CALL CSEND(ITAGS,X,NBYTE,IDS,0)
   CALL CRECV(ITAGR,Y,NBYTE)
ELSE
   CALL CRECV(ITAGR,Y,NBYTE)
   CALL CSEND(ITAGS,X,NBYTE,IDS,0)
ENDIF
```

In this approach, all even numbered nodes send their data to the odd nodes which then receive the data and then they send their data to the even nodes, which are waiting to receive their information. The only requirement being placed on the system is that there is sufficient buffer space to receive a message. This is an environment variable that is beyond the control of the programmer but it is something that should be borne in mind. On the Intel iPSC/860, there is a very large buffer space (which is the reason for the above code samples working) but on the IBM SP2, only 16 KBytes of buffer space is available. For certain programmes, this limit can easily be reached. One remedy is to break up the message into appropriate lengths but care stills needs to be taken to avoid deadlock occurring. A further disadvantage to modifying the message length is that it potentially creates unnecessary work because of a specific architecture. A more sensible approach would be to reconsider the message passing strategy being employed.

13. Communications Involved in Simple Partitioning

This section will now focus on how to solve simple CFD related problems with regular meshes on DM-MIMD machines, such as the Intel or a cluster of workstations. The previous sections have laid the foundations for analysing any results that are obtained. Some simple DO LOOPs will first be investigated to give an indication of what data transfers will result from partitioning a grid. This is typical of what will happen when a finite difference stencil straddles a computational domain.

When a grid is partitioned for allocation to individual processors the complexity of the difference stencil will determine how much data has to be transferred across the subdomains. To illustrate this, we will utilise the High Performance Fortran (HPF) FORALL statement (see Koelbel et al, 1994 for a more complete description) to show the data movement. For ease of presentation we will consider a data set that is allocated to 4 processors with the sequential data set consisting of 16 elements. In this case, processor 1 will have the first 4 elements, processor 2 will have elements 5–8, and so on.

Case 1

```
FORALL (I=1:16) A(I) = B(I)
```

This is a simple assignment requiring only local values and therefore there is no communication involved. It is perfectly parallel.

Case 2

```
FORALL (I=1:15) A(I) = B(I+1)
```

In this case, one element of data has to be transferred in one direction only i.e. A(4) will need to know the value of B(5). The communications involved are illustrated in figure 16a. It is clear that processor 1 does not need to transfer any data and processor 4 does not receive any data. This of course will lead to some imbalance in the work load.

Case 3

```
FORALL (I=2:15) A(I) = B(I+1) - B(I-1)
```

In this case, there will be data transfer in both directions and is more representative of real difference stencils. For this problem, A(5) will need to know the value of B(4). The communications involved are illustrated in figure 16b. If the system were periodic, then processor 1 would also need to transfer data to processor 4.

Case 4

```
FORALL (I=1:15) A(I) = B(INDEX(I))
```

In this case, it is necessary to know the relationship of the values stored in INDEX(I). Without this knowledge, it is impossible to know what the communications requirements are.

Figure 16 Communications Involved in Cases 2 and 3

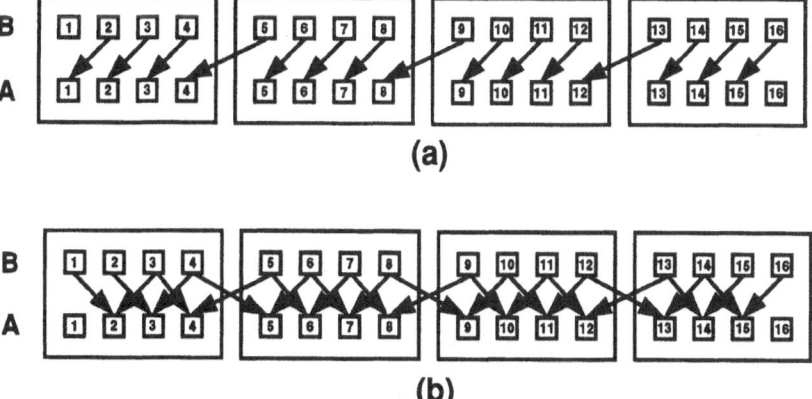

(a)

(b)

14.   A Simple Parallel Example

The technique of partitioning a function is known as *functional* decomposition (Chandy and Taylor, 1992). To illustrate how a program is modified to evaluate a function in parallel, consider a simple integration procedure using the trapezoidal rule. Suppose we wish to evaluate π, then the method would be as follows:

$$\pi = \int_0^1 \frac{4}{1+x^2} dx = \int_0^1 f(x) dx \tag{18}$$

The approach is to divide the area into a number of equispaced intervals, as illustrated in figure 17, and assume each strip forms a rectangle of width W. The value of the

Figure 17   Integration of a function by the Trapezoidal rule

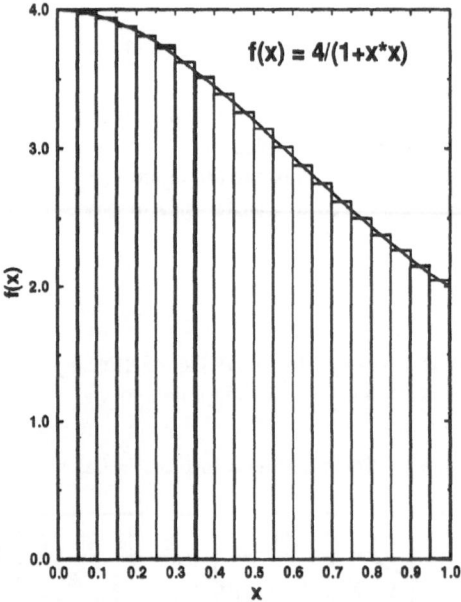

function at the mid-point of the strip is taken to be the height of the rectangle and the area of the rectangle if therefore given by:

$$W \times f(x_{i+1/2}) \tag{19}$$

where W is the width of the interval. The more strips used, the more accurate the calculation will be. A sequential program to integrate the function $f(x)$ could be as follows:

```
PROGRAM TRAP
INTEGER I,N
REAL PI,SUM,W,X
F(X) = 4.0/(1.0 + X*X)
OPEN(UNIT=1,FILE='IN.DAT',STATUS='OLD')
READ(1,*)N
CLOSE(1)
W = 1.0/REAL(N)
SUM = 0.0
DO I = 1 , N
   X = W*(REAL(I) - 0.5)
   SUM = SUM + F(X)
ENDDO
PI = W*SUM
WRITE(*,*)'PI = ',PI
END
```

We now need to modify the program to evaluate the function on a parallel system. Again, the Intel library routines will be used to demonstrate the principles involved. The foundations for setting up any message passing have been covered but this example will be used to illustrate other potential problems that can easily be avoided.

```
PROGRAM TRAP
INTEGER ALLNDS,I,IAM,N,NBYTE,MSGPT0,MSGPT1
REAL PI,SUM,TEMP,W,X
F(X) = 4.0/(1.0 + X*X)
DATA ALNDES,NBYTE,MSGPT0,MSGPT1/-1,4,0,1/
IAM = MYNODE()        ! WHICH NODE AM I
NPROC = NUMNODES()    ! HOW MANY NODES IN CUBE

IF (IAM.EQ.0) THEN    ! ONLY NODE 0 READ IN DATA
   OPEN(UNIT=1,FILE='IN.DAT',STATUS='OLD')
   READ(1,*)N
   CLOSE(1)
   IF (NPROC.GT.1) THEN
      CALL CSEND(MSGPT0,N,NBYTE,ALLNDS,0)
   ENDIF
ELSE
   CALL CRECV(MSGPT0,N,NBYTE)
ENDIF

W = 1.0/REAL(N)
SUM = 0.0
DO I = IAM + 1 , N , NPROC
   X = W*(REAL(I) - 0.5)
   SUM = SUM + F(X)
ENDDO
```

```
IF (IAM.GT.0) THEN    ! ALL NODES SEND DATA TO NODE 0
   CALL CSEND(MSGPT1,SUM,NBYTE,0,0)
ELSE
   DO I = 1 , NPROC - 1
      CALL CRECV(MSGPT1,TEMP,NBYTE)
      SUM = SUM + TEMP
   ENDDO
ENDIF
IF (IAM.EQ.0) THEN ! ONLY NODE 0 HAS THE CORRECT ANSWER
   PI = W*SUM
   WRITE(*,*)'PI = ',PI
ENDIF
END
```

Several new features have now been introduced. The main one is using the CSEND routine to *broadcast* a particular value. A broadcast is an operation which allows a processor to notify all other processors about a particular data item. In this case, the broadcast is being used to inform the other processors how many intervals are to be used. On the Intel, it is done by setting one of the arguments (ALLNDS) to the special value of -1. Many message passing libraries have similar commands. The first major difference between the sequential and parallel program occurs for the I/O. The one thing that should be avoided is all processors trying to access the same data file. It should be noted that this can be handled in a variety of ways and will depend upon the operating system of the architecture being used. It is therefore safer to make it independent of such a feature. The parallel program has only one node accessing the data file and the value that is then read in (i.e. N) is then broadcast to all the other nodes. There is also a built in protection to this section of the code which will prevent node 0 trying to send data to itself, something which can cause a program to hang. This could occur if only one processor is being used (i.e. NPROC = 1).

The program then begins to sum up the individual strips, an operation which is independent of the order and highly parallel. For this example, each processor steps through the loop by a value given by NPROC e.g. if N = 16 and NPROC = 4, then node 0 would add the elements 1, 5, 9 and 13, node 1 would add 2, 6, 10 and 14 etc. Each processor then has a partial value of the global sum and it is therefore necessary to collect this data. The next code section indicates a possible way of doing this. For the example just given, nodes 1, 2 and 3 would send their results back to node 0 and node 0 needs to collect all of these values together. This is done in the DO loop which receives NPROC - 1 values which are stored in the TEMP buffer for the summation to be performed. It should be noted that only node 0 has the final answer. If all nodes needed to know the final answer then a broadcast would be necessary.

The parallel integration routine will work but there is a potential *bottleneck* being caused by all nodes sending their results to node 0, as illustrated in figure 18. The problem is caused by node 0 being unable to process more than one receive at a time i.e. it must process each receive sequentially and node 0 is unable to receive more than one message simultaneously. This potential bottleneck can be removed on the Intel by using the global summation routine GSSUM. The program is modified from:

```
IF (IAM.GT.0) THEN    ! ALL NODES SEND DATA TO NODE 0
   CALL CSEND(MSGPT1,SUM,NBYTE,0,0)
ELSE
   DO I = 1 , NPROC - 1
      CALL CRECV(MSGPT1,TEMP,NBYTE)
      SUM = SUM + TEMP
   ENDDO
ENDIF
```

to the following:

```
CALL GSSUM(SUM,1,TEMP)
```

The 1 in the argument list indicates that SUM is a scalar quantity. The GSSUM routine is provided by Intel and is specifically written to exploit the hypercube architecture.

Figure 18 The bottleneck problem

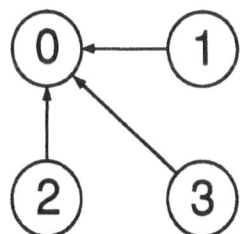

The algorithm used in the GSSUM routine solves the global reduction in $log_2 p$ operations and is very efficient but requires that there are 2^n processors being used. Figure 19 illustrates the principal involved for four processors. At the start of the

Figure 19 The hypercube global reduction algorithm

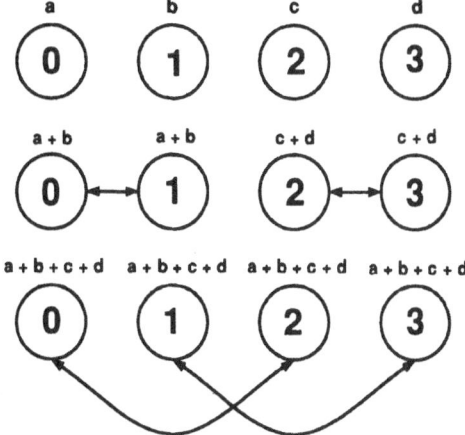

operation, each processor has an initial value. In this example, node 0 knows a, node 1 knows b etc. Nodes 0 and 1 then exchange information and both of these nodes then know the value of a + b. At the same time, nodes 2 and 3 exchange their values and now know c + d. Nodes 0 and 2 then exchange their data and whilst at the same time, nodes 1 and 3 exchange their data. By the end of this exchange, all processors know the value of a + b + c + d. This operation has been completed in 2 steps. This reduction algorithm works exactly the same for vectors i.e. a, b, c and d could have been arrays.

The summing of individual strips, as performed by this program, is a very practical and efficient way to proceed for this particualr problem. However, an alternative way forward is to decompose the problem into groups of strips or "domains" and have each processor perform a summation on its own domain. However, a global communication is still required to obtain the final result. This approach is analogous to a domain decomposition strategy, which will be developed further in the next example, and can be introduced by a simple modification to the start and stop values of the loop counter as follows:

```
IBEG = 1 + IAM*N/NPROC
IEND = (IAM + 1)*N/NPROC
DO I = IBEG , IEND
   X = W*(REAL(I) - 0.5)
   SUM = SUM + F(X)
ENDDO
```

In this approach, IBEG takes the values 1, 5, 9, and 13 and IEND is 4, 8, 12 and 16. However, we have now introduced a potential problem that was not present when the summation was performed using strips. The program now implicitly assumes that N is exactly divisible by NPROC. If it is not, the program will produce an incorrect result. It is left as an exercise to consider what must be done to enable the computation to be performed correctly for any value of N.

As a final look at this parallel example, we will now consider the I/O being performed. It is widely recognised that I/O on parallel systems is much more complicated than on sequential machines. Moreover, I/O can significantly affect the performance of a particular code. To illustrate some of these problems, it is instructive to look at the input for this program. In this version, processor 0 reads in the number of intervals (given by N) and broadcasts the value to all other processors: Is this the best approach? To answer this we will consider what other options are available. Could all processors access the same data file? i.e. can the program be written as the following:

```
READ(1,*)N
CLOSE(1)
```

On the Intel iPSC/860 this approach will work but it will not always work with other operating systems. It also introduces a bottleneck into the computation because each processor is trying to access a single data file. An alternative approach is for each processor to access the data file one at a time. This could be done by the introduction of a simple loop i.e.

```
DO I = 0 , NPROC - 1
   IF (IAM.EQ.I) THEN
      READ(1,*)N
      CLOSE(1)
      CALL GSYNC()
   ELSE
      CALL GSYNC()
   ENDIF
ENDDO
```

However, this introduces a sequential stage into the computation with a forced synchronisation step (given by the GSYNC() routine). This step is necessary because each stage of the read operation will take substantially more time than the time required for each loop iteration and all processors will then be trying to access the same data file, as in the first case. For this particular problem, N could be stored as a parameter and all input would be avoided. The disadvantage of this approach is that the program has to be recompiled every time N is changed. As a final alternative, it is also possible to store NPROC copies of the input data file i.e. IN.DAT0, IN.DAT1 etc. and each processor can then access its own data file, without any conflict. In general, restart files are stored in this way. This allows each processor to output data from its domain to a unique file which can then be used to restart the program for the next set of iterations to be performed.

15. A Simple CFD Example - Poisson's Equation

The most popular approach to solving CFD problems on parallel systems is that of *domain decomposition* although strictly speaking it should be called *grid partitioning*. The objective is to distribute the computational grid onto a number of processors such that the work load on each processor is equivalent. All practicall applications involve the calculation of derivatives. To allow these derivatives to be determined at the partition boundaries it is therefore necessary to include a region of cells around each partition. These are frequently referred to as *halo* cells or *ghost* cells and are illustrated in figure 20. The halo cells need to have data transferred from neighbouring processors, as indicated in the figure, to allow the derivatives to be calculated.

Figure 20 Halo cells arising from grid partitioning

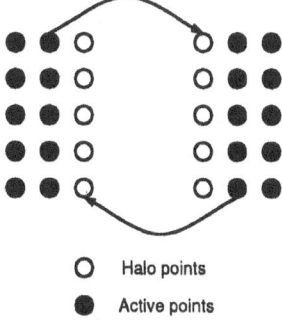

O   Halo points

●   Active points

As an introduction to the technique of partitioning a grid for allocation to individual processors, we will consider a simple two dimensional problem with a constant grid size. The solution of Poisson's equation provides a relatively straightforward example and illustrates the techniques involved for many applications. The Poisson equation can be written as:

$$\nabla^2 T = \frac{\partial^2 T}{\partial x^2} + \frac{\partial^2 T}{\partial y^2} = R(x, y) \qquad (20)$$

and, for the special case when $R(x, y) = 0$, it is known as Laplace's equation. This example will use Laplace's equation to solve a steady state heat conduction problem with the following Dirichlet boundary conditions:

```
y = 0     T = T1     x = 0     T = T2
y = HY    T = T3     x = HX    T = T4
```

as illustrated in figure 21. The boundary temperatures are specified as T1 = 100°C and T1 = T2 = T3 = 0.0°C. The distances HX and HY are taken to be 1m and 2m, respectively, and the grid sizes Δx and Δy are taken to be constant. The initial temperature field is taken to be zero everywhere except at the y = 0 boundary. This

Figure 21 Boundary values for heat conduction problem

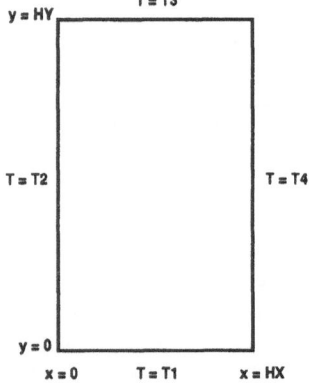

elliptic Partial Differential Equation (PDE) can now be solved, subject to the above conditions, by a variety of methods. For this example, three solution strategies will be considered, namely: (i) Jacobi; (ii) Gauss-Seidel; and, (iii) Conjugate Gradient (without preconditioning). Each approach illustrates different features typically found in CFD. The Jacobi solver is explicit and updates the halo points after each complete iteration. This is typical of many explicit, time-marching schemes used in CFD. The Gauss-Seidel solver makes use of the most recently available data. However, when a grid is partitioned, some of the implicitness is lost and this generally results in more iterations being required to achieve a required convergence criterion. The Conjugate Gradient solver is analogous to many found in CFD. For this particular problem a symmetric set of equations, such as those encountered in pressure correction schemes, are solved and the procedure involves the computation of many dot products. This is a global communication problem and involves sending short messages very frequently.

A discretised form of equation 20 is the following:

$$\frac{T_{i+1,j} - 2T_{i,j} + T_{i-1,j}}{\Delta x^2} + \frac{T_{i,j+1} - 2T_{i,j} + T_{i,j-1}}{\Delta y^2} = R_{i,j} \qquad (21)$$

If we know the initial values throughout the grid, then this can be solved explicitly as follows:

$$T_{i,j}^{n+1} = \alpha\left(T_{i+1,j}^{n} + T_{i-1,j}^{n} + \beta\left(T_{i,j+1}^{n} + T_{i,j-1}^{n}\right) - \Delta x^2 R_{i,j}\right) \qquad (22)$$

where α and β are given by the following expressions:

$$\alpha = \frac{1.0}{2.0(1.0 + \beta)} \qquad (23)$$

$$\beta = \frac{\Delta x^2}{\Delta y^2} \qquad (24)$$

and the superscript n represents the iteration number. This is known as the Jacobi method. It is slow to converge due to the explicit nature of the algorithm. Moreover, because the algorithm is explicit, the Jacobi method takes the same number of iterations to converge for both the sequential and parallel programs.

An improvement to the Jacobi solution technique is to make use of the most recently available values of T. This is known as the Gauss-Seidel method and can be written as:

$$T_{i,j}^{n+1} = \alpha\left(T_{i+1,j}^{n} + T_{i-1,j}^{n+1} + \beta\left(T_{i,j+1}^{n} + T_{i,j-1}^{n+1}\right) - \Delta x^2 R_{i,j}\right) \qquad (25)$$

At the start of the iteration process, the boundary values are known. The solution then sweeps through the x and y planes. However, in the parallel version, some of the sweeps start at an interface boundary and therefore do not have the most up to date values of T. This usually causes some deterioration in the convergence of the solver.

The Conjugate Gradient algorithm is based upon the solution of a system of equations written in the form Ax = b. For this particular case, the matrix A is symmetric and sparse and can be solved by a wide variety of methods. In this example, the psuedocode can be written as:

$$x_0 = 0 \; ; \; r_0 = b - Ax_0$$
$$p_{-1} = 0 \; ; \; \beta_{-1} = 0$$
$$solve \; \omega_0 \; from \; K\omega_0 = r_0$$
$$\rho_0 = (r_0, w_0)$$
$$For \; i = 0, 1, 2, \ldots\ldots$$
$$\quad p_i = \omega_i + \beta_{i-1} p_{i-1}$$
$$\quad q_i = Ap_i$$
$$\quad \alpha_i = \frac{\rho_i}{(p_i, q_i)}$$
$$\quad x_{i+1} = x_i + \alpha_i p_i$$
$$\quad r_{i+1} = r_i - \alpha_i q_i \tag{26}$$
$$\quad If \; \|r_{i+1}\|_2 < \varepsilon \; stop$$
$$\quad \omega_{i+1} = K^{-1} r_{i+1}$$
$$\quad \rho_{i+1} = (r_{i+1}, \omega_{i+1})$$
$$\quad \beta_i = \frac{\rho_{i+1}}{\rho_i}$$
$$end$$

where $(. \; , \; . \; , \; . \;)$ represents a dot product and the tolerance, ϵ, was set to 10^{-6}. In the example chosen, the preconditioning matrix, K, has been set to the identity matrix. In this case, the Conjugate Gradient algorithm performs analagously to the explicit algorithm and converges in the same number of iterations for both the sequential and parallel code. This would not necessarily be the case if a preconditioning matrix were to be used in parallel.

The heat conduction problem selected has an exact solution (Hoffmann, 1989):

$$T = 2T_1 \left[\sum_{n=1}^{\infty} \frac{1 - (-1)^n}{n\pi} \frac{sinh\left(\frac{n\pi(HY-y)}{HX}\right)}{sinh\left(\frac{n\pi HY}{HX}\right)} sin\left(\frac{n\pi x}{HX}\right) \right] \tag{27}$$

The series was truncated at n = 20. It is worth noting that values of x and y are needed throughout the domain and this must be borne in mind when the computational grid is partitioned.

To partition the grid, the following approach has been adopted: the total number of grid points in the x and y directions are specified as IMAXG and JMAXG, respectively. The number of processors in the x and y direction are then specified as NPROCX and NPROCY, respectively. To allow for derivatives at interface boundaries, halo cells need to be added. In this example, only one halo cell is required at each interface. The total number of cells required on a subdomain is therefore IMAX = IMAXG/NPROCX + 2*NHALO in the x-direction and JMAX = JMAXG/NPROCY + 2*NHALO in the y-direction. The foregoing implicitly assumes that there is no remainder from the divisions. If there is a remainder, then some additional work has to be done to redistribute the computational load to try to ensure that the domains do not become too unbalanced. The grid spacing for this problem is then given by $\Delta x = HX/(IMAXG-1)$ and $\Delta y = HY/(JMAXG-1)$, respectively.

Finally, as an indication of how different algorithms perform on a parallel architecture, table 3 shows how a simple program performs on the Intel iPSC/860. Although this is a small problem, the effects of partitioning are very apparent.

Table 3 Results for the Poisson equation

Method	NPROCX	NPROCY	NPROC	ITERS	Time (s)
J	1	1	1	6949	31.74
GS	1	1	1	3642	19.36
CG	1	1	1	150	2.90
J	2	1	2	6949	40.72
GS	2	1	2	3726	20.10
CG	2	1	2	150	2.41
J	1	2	2	6949	25.85
GS	1	2	2	3680	15.57
CG	1	2	2	150	2.63
J	2	2	4	6949	29.46
GS	2	2	4	3763	17.18
CG	2	2	4	150	2.34

16. References

Amdahl, G., "Validity of the Single Processor Approach to Achieving Large Scale Computing Capabilities", In AFIPS Conference Proceedings, 1967, pp 483–485.

Bailey, D., "Twelve Ways to Fool the Masses When Giving Performance Results on Parallel Computers", Supercomputing Review, August 1991, pp 54–55.

Bailey, D., "Misleading Performance in the Supercomputing Field", Supercomputing 92 Proceedings, Published by IEEE, held at Minneapolis, November 16–20, 1992, pp 155–158.

Bailey, D., "RISC Microprocessors and Scientific Computing", RNR Technical Report RNR-93–004, 1993.

Chandi, K. M. and Taylor, S., An Introduction to Parallel Programming, Jones and Bartlett Publishers, Boston, 1992, ISBN 0–86720–208–4.

Dongarra, J. J., Duff, I. S., Sorensen, D. C. and Van der Vorst, H. A., Linear System Solving on Vector and Shared Memory Computers, SIAM, 1991, ISBN 0–89871–270–X.

Dongarra, J. J., Meuer, H. W. and Strohmaier, E., The TOP500 Report, WWW http://parallel.rz.uni-mannheim.de/top500.html

Edmondson, J. H., Rubinfeld, P., Preston, R. and Rajagopalan, V., "Superscalar Instruction Execution in the 21164 Alpha Microprocessor", IEEE Micro, April 1995, pp 33 – 43.

Emerson, D. R., Blake, R. J. and Allan, R. J.,"Unsteady Flow Over Multiple Cavities on MIMD and Virtual Shared Memory Architectures", in CFD Algorithms and Applications for Parallel Processors, ASME Fluids Engineering Conference FED-Vol 156, June 20–24, 1993, Washington, pp 33–41.

Flynn, M. J., "Some Computer Organisations and Their Effectiveness", IEEE Transactions on Computers, Vol. C-21, No. 9, Sept. 1972, pp 948–960.

Gajski, D. D. and Pier, J-K, "A Comparison of Five Multiprocessor Systems", Parallel Computing, Vol 2, 1985, pp 265–282.

Gustafson, J. L., "Reevaluating Amdahl's Law", Communications of the ACM, Vol 31, No. 5, 1988, pp 532–533.

Heath, M. and Worley, P., "Once Again, Amdahl's Law", Communications of the ACM, Vol 32, No. 2, February 1989, pp 262–264.

Hill, M. D., "What is Scalabilty?", Computer Architecture News, December, 1990, pp 18–21.

Hockney, R. W. and Jesshope, C. R., Parallel Computers 2, IOP Publishing, 1988, ISBN 0–85274–812–4.

Hoffmann, K. A., Computational Fluid Dynamics for Engineers, Engineering Education System, 1989, ISBN 0–9623731–4–1.

Hord, R. M., The Illiac IV - The First Supercomputer, Springer-Verlag, 1982, ISBN 0–387–11765–2.

Hwang, K. and Briggs, F. A., Computer Architecture and Parallel Processing, McGraw-Hill, 1985, ISBN 0–07–031556–6.

Koelbel, C. H., Loveman, D. B., Schreiber, R. S., Steele, G. L. and Zosel, M. E., The High Performance Fortran Handbook, MIT Press, 1994, ISBN 0–262–11185–3.

Kuck, D. J., The Structure of Computers and Computations, John Wiley and Sons, 1978, ISBN 0–471–02716–2.

Lawrie, D. H., "Access and Alignment of Data in an Array Processor", IEEE Trans. on Computers, Vol C-24, No. 12, December 1975, pp 1145–1155.

Leiserson, C. E., "Fat-Trees: Universal Networks for Hardware-Efficient Supercomputing", IEEE Trans. on Computers, Vol C-34, No. 10, October 1985, pp 892–901.

Menabrea, L. F., "Sketch of the Analytical Engine Invented by Charles Babbage", ESP. Bibliothèque Universelle de Genève, 82, 1842.

Mirapuri, S., Woodacre, M. and Vasseghi, N., "The MIPS R4000 Processor", IEEE Micro, April 1992, pp 10 – 22.

Nussbaum, D. and Agarwal, A., "Scalabilty of Parallel Machines", Communications of the ACM, Vol 34, No. 3, March 1991, pp 57–61.

Oed, W.,"The Cray Research Massively Parallel Processor System: CRAY T3D", Cray Research, November, 1993.

Pease, M. C., "The Indirect Binary n-Cube Microprocessor Array", IEEE Trans. on Computers, Vol C-26, No. 5, May 1977, pp 458–473.

Randell, B., The Origins of Digital Computers, Springer-Verlag, 1982, ISBN 3–540–11319–3.

Singh, J. P., Hennessy, J. L. and Gupta, A., "Scaling Parallel Programs for Multiprocessors: Methodology and Examples", IEEE Computer, July, 1993, pp42–50.

Trew, A. and Wilson, G. (Eds.), Past, Present, Parallel - A Survey of Available Parallel Computing Systems, Springer-Verlag, 1991, ISBN 0–387–19664–1.

Ware, W., "The Ultimate Computer", IEEE Spectrum, March 1973, pp 89–91.

PROGRAMMING ASPECTS AND ALGORITHMS FOR VECTOR- AND PARALLEL COMPUTERS

C. LACOR
Professor
Vrije Universiteit Brussel
Dept. Fluid Mechanics
Pleinlaan 2
1050 Brussel

1. Programming aspects

1.1 INTRODUCTION

The development of faster and faster computers can be based on two different techniques.

The first is to increase the speed of the processor by improving the hardware such that the clock cycle time decreases. With current computer technology the limit of what physically can be achieved is being approximated. As a consequence an increase of computing power with an order of magnitude is not to be expected anymore from this approach.

Another way is to built in some form of concurrency i.e. the simultaneous operation of one or more hardware functions. An historical overview of the use of concurrency and its applications in computers can be found in Gentsch and Neves (1988) and Hockney and Jesshope (1988).

Concurrency is used in current day computers under many forms.

An obvious form is replication like for instance the replication of CPUs leading to parallel computers.

Another way is pipelining on which principle vectorcomputers are based.

In this first part of the lecture, vectorcomputers and parallel computers will be discussed.

Note that some computer constructors combine both approaches to concurrency in their computers.

43

P. Wesseling (ed.), High Performance Computing in Fluid Dynamics, 43–95.
© 1996 *Kluwer Academic Publishers.*

1.2 VECTORCOMPUTERS

As mentioned above vectorization is based on pipelining.

In the present section the most important issues of vectorcomputers are discussed first: pipelined functional units, Amdahl's law, chaining, paths to the memory and memory conflicts.

Next some general rules for vectorization will be discussed, as well as some general optimization techniques and the optimization of basic vector operations.

1.2.1 Pipelining

Before discussing pipelining, first consider the addition of two scalar variables A and B on a **scalar machine**, figure 1.2.1.1.

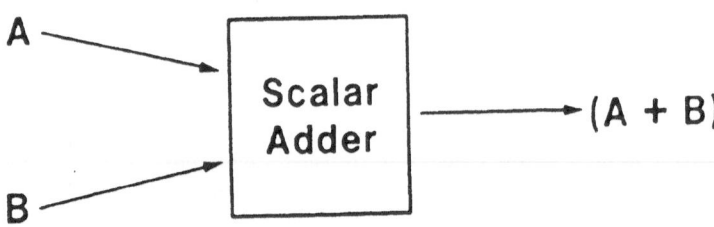

Figure 1.2.1.1 : Symbolic representation of scalar addition.

Assume that this operation is performed in n clock cycles.

If A and B were vectors of length N, the addition of all their components, would require N*n clock cycles. If the cycle time is t, the total time T_s (where the subscript stands for scalar) is given by

$$T_s = N*n*t \qquad\qquad (2.1)$$

In a pipelined machine the functional units are especially designed for operations on vectors.

They are conceived as automobile assembly lines, figure 1.2.1.2.

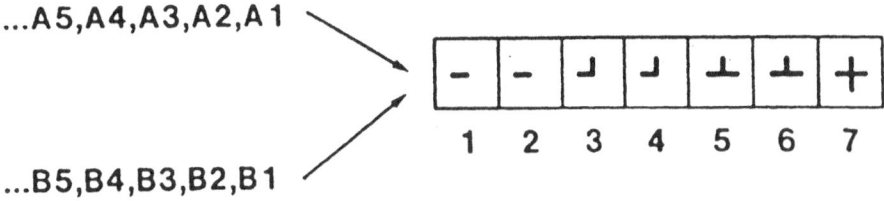

Figure 1.2.1.2 : Symbolic representation of pipelined addition.

The functional unit, symbolically shown in figure 1.2.1.2, consists of 7 segments. If two scalars A and B are to be added the result will be available after 7 clock cycles.

If however two vectors are to be added the 'pipelining mechanism' comes into play. This can be understood by an assembly analogy.

As soon as the first components of the vectors, A(1) and B(1), enter segment number 2, the second components, A(2) and B(2), are fed into segment number 1. Similarly if A(1) and B(1) move to segment number 3, A(2) and B(2) enter segment number 2 and A(3),B(3) enter segment number 1. This procedure goes on until after 8 clock cycles the first result, i.e. A(1)+B(1) becomes available. From then on, a new result will be generated every clock cycle, i.e. A(2)+B(2) after 9 cycles, A(3)+B(3) after 10 cycles and so on.

If the functional unit has m segments, the addition of vectors A and B (of length N) requires a total time T_v (subscript v stands for vector) given by

$$T_v = (m + N)*t \qquad (2.2)$$

The number m*t is the time that elapses after the first pair of data entered the pipeline and the first result leaves the pipeline. This is also called the **startup time** s.

The **performance gain** G on the pipelined machine can be defined as the ratio of T_s and T_v :

$$G = \frac{n}{1 + \frac{m}{N}} \qquad (2.3)$$

From equ. (2.3) it is clear that pipelining leads to a speed up provided :

$$N > \frac{m}{n-1} \qquad (2.4)$$

i.e. the vectors need to have a certain minimum length. This minimum vector length is referred to as the **crossover point**.

In the previous example, if the scalar addition requires 4 cycles (n=4) and the pipelined adder has 7 segments (m=7), the crossover point corresponds to a length of 3 (i.e. N > 7/3).

On Cray machines DO loops with only one or two iterations will therefore not be vectorized by the compiler. Vectorization is attempted for loops with a trip count of 3 or more or with an unknown trip count.

Note for a given n and m, the maximum speed up is obtained in the limit of $N \rightarrow \infty$, and is given by n.

The **performance P** of any computer is usually expressed in million floating point operations per second,(MFLOPS).

The performance obtained for the addition of the matrices A and B of our example, can be calculated from equation (2.2). One finds :

$$P = \frac{1}{t + \dfrac{s}{N}}. \, 10^{-6} \, \text{MFLOPS} \qquad\qquad (2.5)$$

where, as above, t is the clock cycle and s the startup time.

Equation (2.5) leads to the curve of figure 1.2.1.3 in the P/N plane.

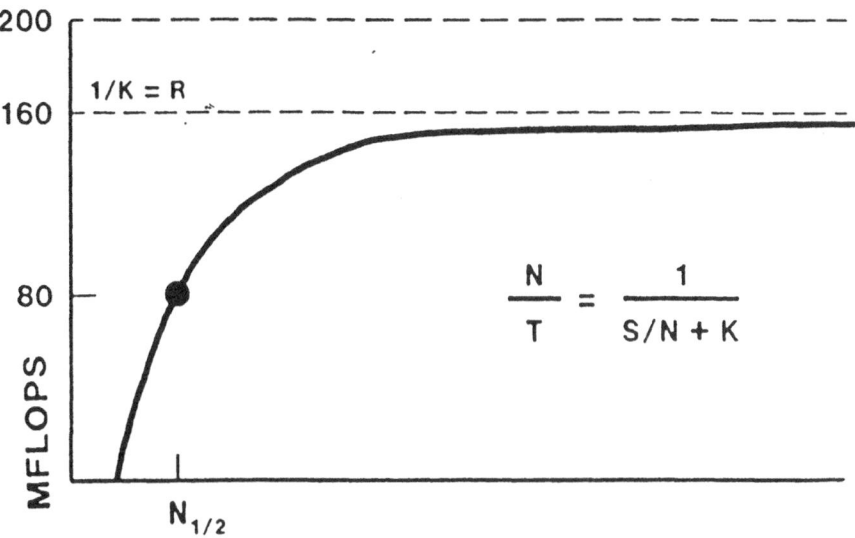

Figure 1.2.1.3 : Characteristic curve for a vector operation.

The maximum performance is $10^{-6}/t$ and is obtained for $N \to \infty$. This number is also referred to as the **peak performance P_{max}**.

Consider for instance the CRAY X-MP/14 with a clock cycle period of $9.5 \cdot 10^{-9}$ seconds. The peak performance according to equ. (2.5) is $10^3/9.5$ MFLOPS, i.e. ± 105 MFLOPS.

However the actual peak performance of this machine is 210 MFLOPS. This doubling of the performance is caused by **chaining**, a technique to be discussed in the next section.

On parallel computers with multiple CPUs the peak performance is obtained by multiplying the peak performance of 1 CPU with the total number of CPUs.

Though the timing model used above, as expressed by equ. (2.2), is accurate enough for the present purposes, it can still be fine tuned.

Bucher (1983) gives the following more accurate timing model taking the length of vectorregisters into account. These registers are usually 64 words long. Therefore, if e.g. a vector of length 250 is to be processed, it is split up into 3 vectors of length 64 and one vector of length 58.

$$T_V = s_{out} + (\frac{s_{in}}{64} + t) N \qquad (2.6)$$

where s_{out} is the initial startup time for the whole process (the outer loop), s_{in} is the startup time of the inner loop of length 64 or less that can not be overlapped with other vector instructions.

The peak performance alone is insufficient to characterize the performance of a supercomputer.

It is an asymptotic performance rate and should rather be considered as the performance that the computer is guaranteed not to exceed

1.2.2 Amdahl's law

This law models the **performance gain** G or the **speed-up ratio** on vectorcomputers as a function of the ratio of vector and scalar speed and of the percentage of the code that has been vectorized.

Note that an expression for G was already obtained in the previous section. There the ratio of vector and scalar speed was modeled using machine characteristics (n,m) and it was assumed that the application was fully vectorized.

Suppose now that in a certain application N floating point operations have to be performed. Further assume that S is the number of floating point operations that can be executed per second in Scalar mode, and V the number of floating point operations that can be executed per second in Vector mode.

Suppose that a fraction f of the total number of operations is executed in vector mode. The total execution time T_V is then

$$T_V = \frac{f*N}{V} + \frac{(1-f)*N}{S} \qquad (2.7)$$

If all operations were performed in scalar mode the total execution time T_S would be

$$T_S = \frac{N}{S} \qquad (2.8)$$

The performance gain G obtained by using vectorization, is then given by

$$G = \frac{1}{(1-f)+\frac{f.S}{V}} \qquad (2.9)$$

The first conclusion from equation (2.9) is that - even if $V/S \rightarrow \infty$ - the maximum performance gain G_{max} is

$$G_{max} = \frac{1}{1-f} \qquad\qquad (2.10)$$

Figure 1.2.2.1 illustrates the gain in function of the vectorized percentage both for an infinitely fast vector speed and for a vector speed which is ten times the scalar speed.

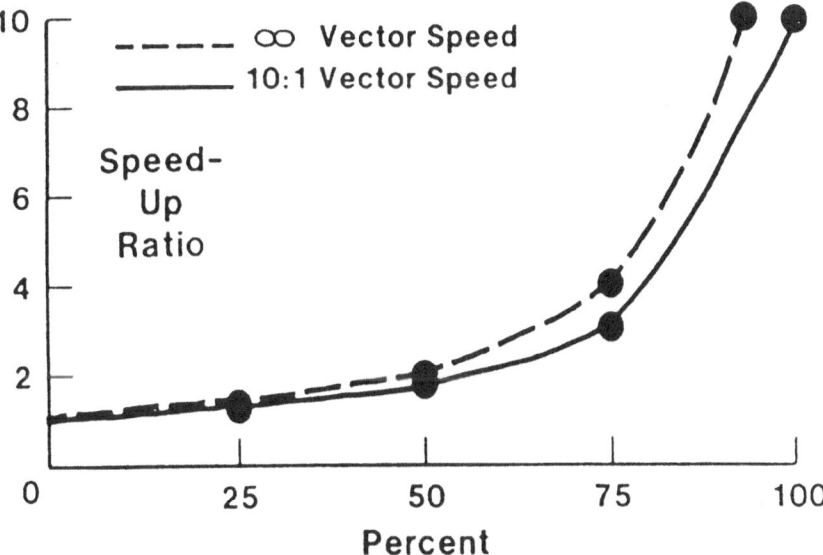

Figure 1.2.2.1 : Illustration of Amdahl's law for two different vector speeds.

From eq.(2.5) and fig.1.2.2.1 it is seen that in order to take full advantage of the speed of vector processing a high percentage of the code has to be vectorized. Even for a vectorization percentage of 90, the maximum gain is only 10.

Further, assume that 75% of the code is vectorized. If one could increase the vector speed from 10 times scalar speed to infinite speed - which would be a tremendous hardware improvement - the overall performance would rise from 3.63 to 4 times scalar speed.

If we assume however that instead of this hardware improvement we would improve the software, such that 90% of the code is vectorized (which may not be too difficult), the performance rises to 5.26 times scalar speed.

This example illustrates that architectural improvements are only **opportunities** for performance improvements and that the opportunity is **not realized** except in software that can take advantage of the hardware design.

1.2.3. Chaining

Chaining, also called **linking**, is a technique whereby different pipelined processes are successively combined.

This means that if for instance 2 pipelined units have to be used successively, a result leaving the first pipeline will immediately be fed into the second pipeline, without having to wait for the first pipeline to have processed all the data.

Chaining is best illustrated with an example. Consider the following DO loop :

```
    DO 10 I = 1,N
        A(I) = X(I)*Y(I) + Z(I)
 10 CONTINUE
```

The process of chaining is visualized in figure 1.2.3.1 :

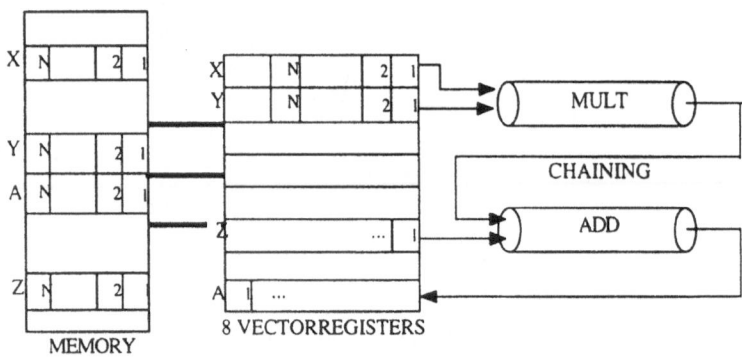

Figure 1.2.3.1 : The principle of chaining on a vectormachine.

As soon as the first result of the product X(I)*Y(I) leaves the MULT pipeline, it is fed into the ADD pipeline.

Let's investigate the advantage of chaining into some more detail.

If no chaining is available, the processing of this DO loop can be visualized as shown in figure 1.2.3.2. For simplicity it is assumed that all the vectors are already in the vector registers.

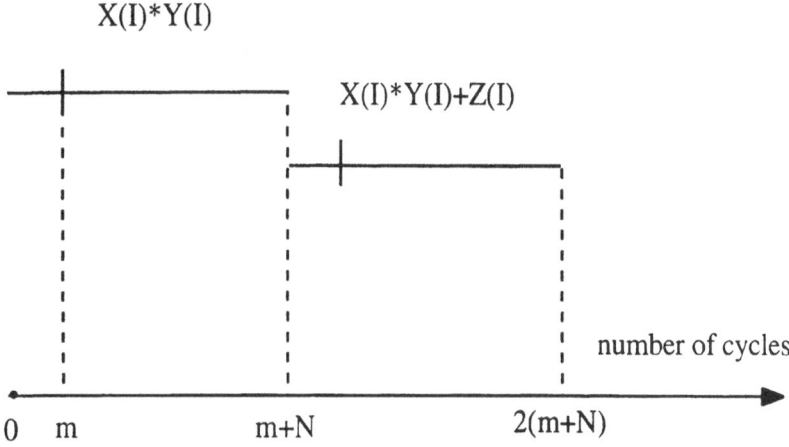

Figure 1.2.3.2 : Schematic representation of the processing of vector operations without chaining.

The horizontal lines represent the processing of the vector operations in time (expressed in numbers of clock cycles).

Each operation consists of a startup time (first m cycles, as indicated by the small vertical bar) followed by N cycles to process the N vector elements.

In figure 1.2.3.2, the addition of the vector Z can not start before the operation $X(I)*Y(I)$ is completed. The performance for the above DO loop, without chaining, is then found as follows :

> 2N floating point operations (multiplication and addition) in $2(m+N)t$ seconds

This leads to the performance of equation (2.5).

The same operation with chaining is sketched in figure 1.2.3.3.

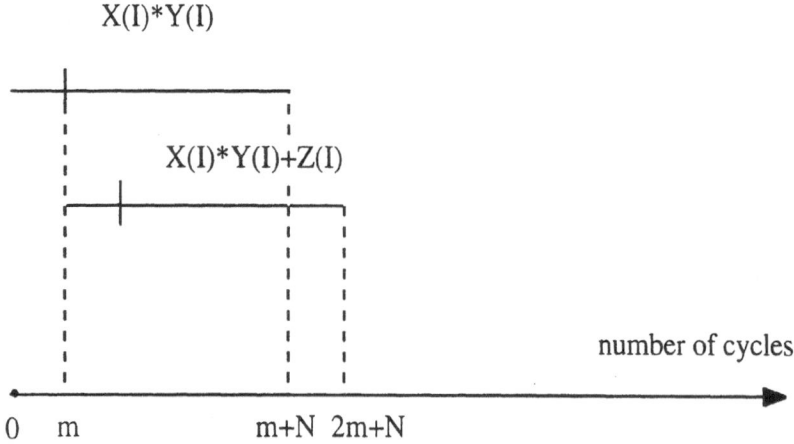

Figure 1.2.3.3 : Schematic representation of the processing of vector operations with chaining.

As soon as the first result, in this example X(1)*Y(1), leaves the pipeline it enters the second pipeline. This leads to the following balance :

2N floating point operations in (2m+N)t seconds

or a performance given by :

$$P = \frac{2}{t + \dfrac{t.2m}{N}}.10^{-6} \qquad (2.11)$$

If this result is compared with the result of equation (2.5) it is clear that the performance is doubled as a result of the use of chaining.

1.2.4. Paths to Memory

The flow of data from the memory to the computational units is the most critical part of the computer design.

The aim is to keep the functional units running at their peak capacity. To this end, one uses an **interface** between the central memory and the vector units.

Most supercomputers use **vector registers** as interface. The vector functional units operate only on operands that are stored in these vector registers. These vector registers are directly connected to the memory. These connections are also referred to as **paths to memory** or **ports**.

The number of these connections varies from machine to machine. It will be seen below that the number of paths to memory has a large influence on the computer performance.

The effect of the number of paths to memory on the performance is discussed with the so-called **SAXPY** operation as an example :

```
DO 10 I = 1,N
   Y(I) = A*X(I) + Y(I)
10 CONTINUE
```

The performance of this operation is first evaluated on a machine with 1 path to memory.

Figure 1.2.4.1 illustrates this operation, using the same representation as in the section concerning chaining.

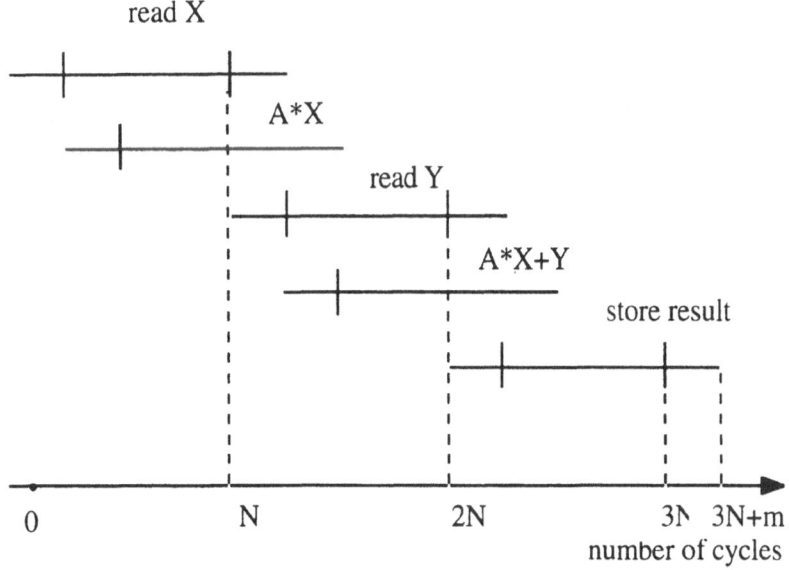

Figure 1.2.4.1 : Processing of SAXPY operation on machine with chaining and 1 path to memory.

First the vector X must be fetched from memory to the vector register. The vector Y can not be accessed from memory at the same time, since there is only 1 path to the memory.

Note that the process of **memory access** is also **pipelined**. As a result, during a vector fetch from memory, some machine cycles elapse before the first element is in the register. This so-called **latency** is similar to the startup time of the vector functional units. For simplicity, we will assume that the corresponding number of cycles is the same as for a vector operation.

The latency is indicated in figure 1.2.4.1 by a vertical bar. Note that a memory access in figure 1.2.4.1 is characterized by 2 vertical bars. The first bar indicates the time that the first element enters the register, whereas the second one corresponds to the time that the last element enters the pipeline. The vector fetch ends if the last element enters the register.

Provided chaining is available, the multiplication with the scalar A can start as soon as the first value of X is in the register, cf. figure 1.2.4.1.

As soon as the last element of X is in the pipeline to the registers (2nd vertical bar), the reading of vector Y can start.

Once Y(1) is available the addition operation can start immediately.

The storage of the final results back to memory can only start when the last element of Y is in the pipeline to the registers.

Inspection of figure 1.2.4.1 shows that the 2N floating point operations are performed in 3N+m cycles (assuming m cycles latency and startup).

The peak performance (for high values of N) is therefore only 2/3 of the maximum as predicted by equation (2.5), i.e.

$$P_{max} = \frac{2}{3t} \cdot 10^{-6} \qquad\qquad (2.12)$$

Consider now a machine with 2 paths to memory; the processing scheme is shown in figure 1.2.4.2.

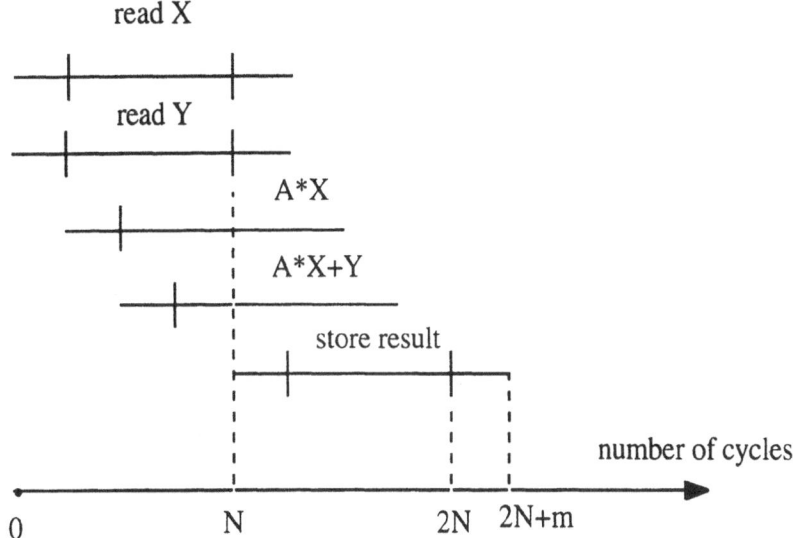

Figure 1.2.4.2 : Processing of SAXPY operation on machine with chaining and 2 paths to memory.

Vectors X and Y can now be fetched simultaneously.

The multiply can start as soon as X(1) is available and the addition with Y can start almost immediately after.

The only 'dead time' occurs when storing the result. The machine has to wait until X(N) is read from memory.

The 2N floating point operations are performed in 2N+m cycles, leading to a maximum performance of

$$P_{max} = \frac{1}{t} \cdot 10^{-6} \qquad\qquad (2.13)$$

The processing on a machine with 3 paths to memory is shown in figure 1.2.4.3.

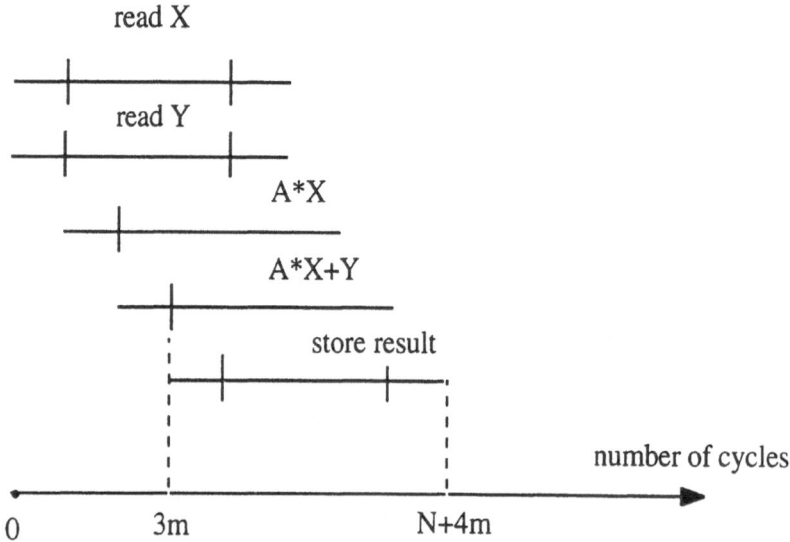

Figure 1.2.4.3 : Processing of SAXPY operation on machine with chaining and 3 paths to memory.

The maximum performance is now

$$P_{max} = \frac{2}{t} \cdot 10^{-6} \qquad (2.14)$$

Comparison of equations (2.12) and (2.14) shows that, without any increase in the speed of the functional units, a factor of 3 in loop performance is achieved.

Yet, supposing that the 1-path and the 3-path machine use the same chip technology and functional unit structure, the manufacturers of both machines would give the same peak performance rates !

1.2.5 Memory Conflicts

1.2.5.1. Memory organization and Conflicts Most main memories of computers have a so-called **memory cycle time.** This is the time required between successive data fetches. If the memory is accessed to retrieve a piece of data, it must 'recover' some cycles before it can be accessed again.
In order to reduce the unfavorable effects of this memory cycle time on the performance, the memory is grouped in '**banks**' or logical partitions.

Each bank has again a cycle time, but different banks can be accessed continuously, i.e. without cycle time.

In order to use the banks in a proper way, **contiguous** sets of data, such as vectors and arrays, are stored in memory in an **interleaved** way.

This means that **consecutive elements** of arrays and vectors are stored in **consecutive banks**.

This arrangement allows to avoid the effects of the cycle time in a vector or array fetch, provided the fetched data does not come from a very recently accessed bank.

In the latter case, the bank possibly has not recovered yet, and one must wait some clock cycles before it can be accessed. One says that there is a **bank busy conflict**.

Figure 1.2.5.1 shows a CPU with 3 paths to the memory (A,B,C) and 16 banks, grouped in 4 **sections. Consecutive banks** are put in **different sections**. As a result, consecutive elements of A are not only put in different banks but also in different sections.

Each of these sections has an independent **path** to the CPU, as illustrated in figure 1.2.5.1.

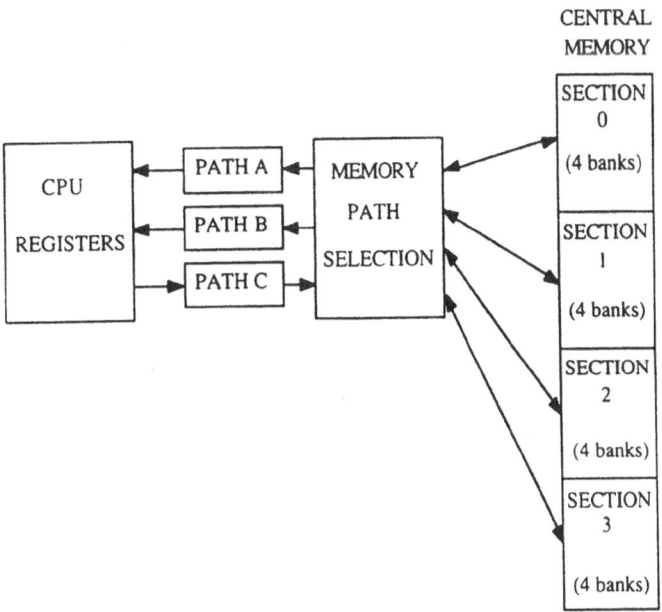

Figure 2.2.5.1 : Organization of the memory on a CRAY X-MP.

These paths to the sections are not to be confused with the paths to memory, described in section 2.4.

The paths to memory are denoted PATH A to C in figure 2.2.5.1 : indeed the CRAY X-MP has 3 such paths, of which two for data retrieval (paths A and B) and one for data storage (path C).

In a multiprocessor machine each CPU is structured in the same way.

The CPU has **four independent paths to the four sections**. This allows the CPU to make up 4 simultaneous memory references, one reference to each section.

However the highest rate for retrieval or storage of a single vector is only one element per clock period. This is because only a single path to the memory can be used for that vector.

The independent paths to the sections are useful when two different vectors are to be fetched, as illustrated below.

Consider the following loop :

```
    DO 10 I = 1,100
       Z(I) = X(I) + Y(I)
 10 CONTINUE
```

The two vectors X and Y in this loop have to be loaded in the registers.

If the corresponding elements of X and Y - i.e. X(I), Y(I) - are in different sections, they can be loaded simultaneously. Since there are also 2 paths to the memory, for loading vectors in the registers, the two vectors can be stored in the registers in the same time as a single vector.

If however X and Y have corresponding elements in the same section a **section conflict** results : the paths A and B try to access the same memory section, but there is only one path to that section ! Note that the programmer has no control or influence on this type of conflict.

The section conflict is **resolved by giving priorities** to the paths A,B and C. The path with the highest priority is allowed to access the section and the other path has to wait one clock period.

Another possible conflict is the **bank busy conflict**, already mentioned above.

This occurs when one tries to access a memory bank before it has completed a previous access. This conflict usually costs more than one clock cycle and is therefore more severe than the section conflict.

On CRAY X-MPs with more than one processor another conflict exists : the **simultaneous bank conflict**.

This occurs when two or more of the paths to the memory, belonging to **different CPUs** try to simultaneously access the **same bank**. Note that no section conflict occurs as each CPU has its own path to the section.

This conflict is **resolved by a rotating CPU priority** system. The CPU currently holding the highest priority is allowed to access the bank, the others must wait one clock period.

This conflict is always followed by a bank busy conflict !

How can bank busy conflicts occur ?

One possibility is when going through a vector in *non-consecutive positions*, as in do loops with a stride.

Consider the following loop with stride 12:

```
      DO 10 I = 1,100,12
         C(I) = A(I) + B(I)
   10 CONTINUE
```

Figure 1.2.5.2 illustrates the retrieval of vector A from memory in the loop above, assuming there are 16 banks.

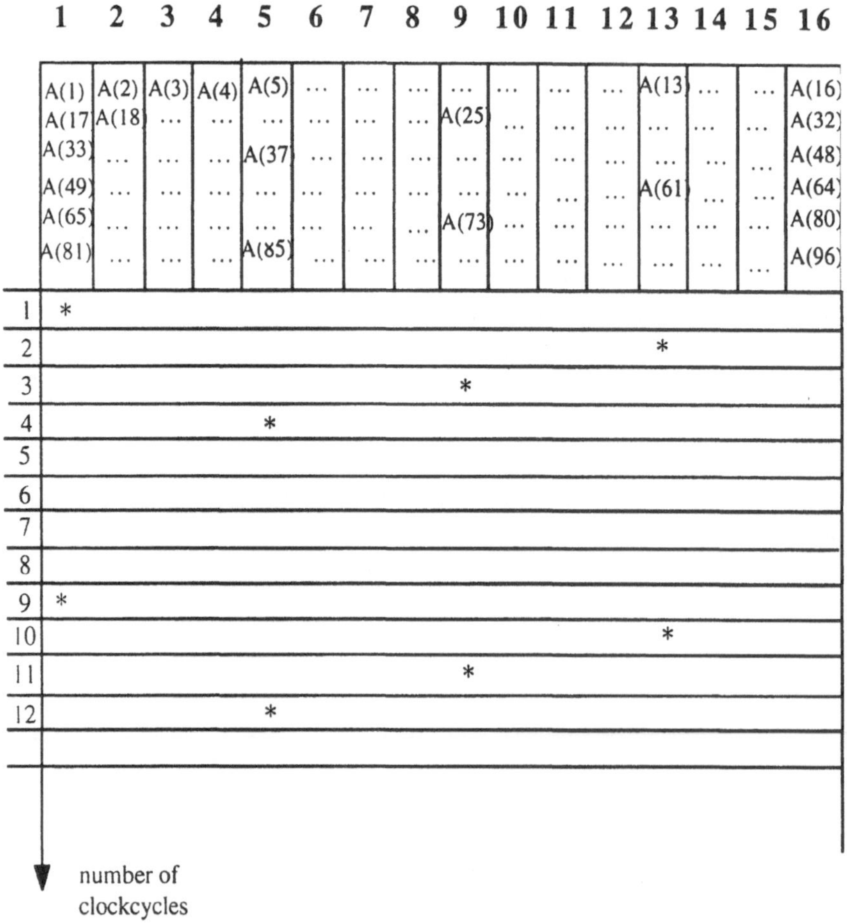

	1	2	3	4	5	6	7	8	9	10	11	12	13	14	15	16
	A(1)	A(2)	A(3)	A(4)	A(5)	...	...	...	...	...	...	...	A(13)	...	...	A(16)
	A(17)	A(18)	...	...	...	...	...	...	A(25)	...	...	...	...	...	...	A(32)
	A(33)	...	...	...	A(37)	...	...	...	...	...	...	...	...	...	...	A(48)
	A(49)	...	...	...	...	...	...	...	...	...	...	A(61)	...	...	...	A(64)
	A(65)	...	...	...	...	...	...	...	A(73)	...	...	...	...	...	...	A(80)
	A(81)	...	...	...	A(85)	...	...	...	...	...	...	...	...	...	...	A(96)

cycle	1	2	3	4	5	6	7	8	9	10	11	12	13	14	15	16
1	*															
2													*			
3									*							
4					*											
5																
6																
7																
8																
9	*															
10													*			
11									*							
12					*											

number of clockcycles

Figure 1.2.5.2 : Illustration of performance degradation as a result of bank busy conflicts.

On top of the figure the distribution of the elements of A along the 16 different banks is shown. On the vertical axis the retrieval of A is sketched as a function of the number of clock periods. The access of a bank is indicated by a * .

In the loop above the elements A(1),A(13),A(25),A(37),A(49),A(61),A(73),A(85),A(97) have to be loaded consecutively. One says that the elements are accessed with a **stride** of 12.

The first 4 elements are in different banks and are loaded in 4 clock periods, cf. figure 1.2.5.2.

A(49) however is in the same bank as A(1). Assuming that the bank latency time is 8 clock periods (CRAY X-MP/14), A(49) can not be loaded on clock cycle 5. Instead one has to wait until clock cycle 9, cf. figure 2.17.

The next 3 elements of A - A(61),A(73),A(85) - can be loaded within 3 clock cycles.

The loading of A(97) results again in a bank conflict (not shown anymore on the figure).

From figure 1.2.5.2 the loss in performance as a result of this bank conflict can easily be calculated.

It is seen that one obtains 4n results in (n-1)*8 + 4 clock cycles. Compared to the full performance, i.e. one result per clock cycle, in the limit of n infinite, the **performance drops by a factor of 2**. Other strides may lead to a more severe degradation.

Another common occurrence of bank conflicts is when accessing *multidimensional arrays*.

This is especially true when the *innermost loop* is *not on the first array index*, like in the example below

```
      DIMENSION A(32,50),B(32,50)
      . . .
      DO 10 I = 1,32
        DO 10 J = 1,50
          B(I,J) = A(I,J) * T1
   10 CONTINUE
```

Bank conflicts can now occur because the elements of multidimensional matrices are stored columnwise in Fortran (the first index is varied first).

It is easily seen that in the DO loop above the same bank is accessed for the retrieval of the elements of A in the innermost loop.

Note that if the DO loop is rewritten as :

```
      DIMENSION A(32,50),B(32,50)
      . . .
      DO 10 J = 1,50
        DO 10 I = 1,32
          B(I,J) = A(I,J) * T1
   10 CONTINUE
```

i.e. the **loops are inverted** (the indices of innermost and outermost loop are switched), **no bank busy conflicts** occur, and the performance is increased with a factor of 8.

1.2.5.2. Techniques to avoid bank busy Conflicts. A simple technique to remove bank busy conflicts is **loop inversion**, cf. the last example of 1.2.5.1.
Another technique consists in transposing the matrix, i.e. **switching** its **dimensions**. This is only effective if the other dimension is not a multiple of the number of banks.

Finally one can also **extend the dimensions**. The dimension is artificially extended so that it is not a multiple anymore of the number of banks.

1.2.6 Vectorization

Only **innermost loops** are candidates for vectorization. Unfortunately, not all innermost loops are vectorizable.

One possible reason is the occurrence of READ, WRITE statements in the loop, or the occurrence of calls of other subroutines

Another possibility is the occurrence of dependencies leading to different results in vector and scalar processing, cf. below..

Some of these unvectorizable loops may be made safe for vectorization by some simple changes, as will be seen in subsequent parts of this course.

1.2.6.1. Dependencies.   Consider the following loop :

```
      DO 10 I = 2,3
        B(I) = A(I-1) + D(I)        ! s1
        A(I) = C(I)          ! s2
   10 CONTINUE
```

Using uppercase letters to indicate initial values and lowercase for new values, the loop above can be written out on a scalar machine as :

```
   b(2)  = A(1)  + D(2)
   a(2)  = C(2)
   b(3)  = a(2)  + D(3)
   a(3)  = C(3)
```

In the loop above there is a dependency as the element a(2) is first defined (i.e. appears on the left-hand-side) in statement s2, and later on (here the next iteration) the same element is referenced (i.e. appears on the right-hand-side) in statement s1. This is a so-called **flow dependency**.

Let's look at the execution of this loop on a vector machine. Pipelined functional units are used now, such that the first statement (s1) *will be executed for the complete vector* before going to the next statement.

```
   b(2)  = A(1)  + D(2)
```

```
b(3) = A(2) + D(3)
a(2) = C(2)
a(3) = C(3)
```

It is clear that this loop leads to different results when executed in scalar and in vector mode. The result is that the compiler will not vectorize the loop.

This does not mean that all flow dependencies inhibit vectorization. If the order of the statements is switched in the loop above, i.e.

```
DO 10 I = 2,3
   A(I) = C(I)           ! s1
   B(I) = A(I-1) + D(I)       ! s2
10 CONTINUE
```

there is still a flow dependency, but one can easily verify that the loop is vectorizable.

An example with another type of dependency is

```
DO 10 I = 1,2
   A(I) = C(I) + D(I)         ! s1
   B(I) = A(I+1)     ! s2
10 CONTINUE
```

Executed in scalar and vector mode this loop can be expanded as :

Scalar mode :

```
a(1) = C(1) + D(1)
b(1) = A(2)
a(2) = C(2) + D(2)
b(2) = A(3)
```

Vector mode :

```
a(1) = C(1) + D(1)
a(2) = C(2) + D(2)
b(1) = a(2)
b(2) = A(3)
```

The dependency is again on vector A : A(2) is first referenced (in s2) and later defined (here the next iteration in s1). This is an **anti-dependency**, which, in this example, inhibits vectorization.

Note that in the example above, the loop, executed in scalar mode, remains unchanged if the order of the statements s1 and s2 is switched. In this case however the loop also becomes vectorizable.

A last example is :

```
DO 10 I = 2,3
   A(I) = C(I) + D(I)       ! s1
   A(I-1) = B(I)     ! s2
10 CONTINUE
```

Processed in scalar mode the loop reads :

```
a(2) = C(2) + D(2)
a(1) = B(2)
a(3) = C(3) + D(3)
a(2) = B(3)
```

The dependency is again on A, but now a(2) is defined in s1 and again defined later (next iteration in s2). This is an **output dependency**.

It is clear that this loop can be vectorized. However note that the same loop with the statements in the reverse order can not be vectorized anymore.

Remark

Some dependencies, which at first glance inhibit vectorization, may still allow a (partial) vectorization. This is the case when the dependency has a **treshhold.** The treshhold is the number of iterations that occur before a value is reused.

Example with flow dependency:

```
DO 10 I = 10,100
   B(I) = A(I-9) + D(I)            ! s1
   A(I) = C(I)             ! s2
10 CONTINUE
```

The treshhold in the example above is 9. Expansion of the loop shows that it can be vectorized with a vector length of 9 : if one executes the statements in blocks of 9 iterations in vector mode, the result is identical to scalar processing.

1.2.6.2 Conditions that prevent vectorization. Apart from dependencies, the following conditions can also prevent vectorization of an innermost loop :
   • An I/O statement.

   • Any CALL of subroutine and any reference to an external function or an intrinsic function that is not vectorizable. Also the use of RETURN, STOP or PAUSE within the loop.

   • IF statements; though some do not prevent vectorization it is always better to avoid them inside a loop, since even with vectorization the performance of the loop will significantly be lower than that of a vectorized loop without IF statement.

   • Backward branches (other than the one that forms the loop).

   • Statements that branch into the loop from outside the loop.

Also note that too much complexity inside the loop can also prevent vectorization, because the required analysis is judged too demanding of system resources, relative to the performance improvement that is anticipated in the code.

1.2.7 Optimization Techniques

1.2.7.1 Loop Unrolling. The technique of **loop unrolling** can be applied to **nested** loops.
It consists in **expanding out** (some of) the iterations of a loop level.

In **vertical** unrolling the expansion increases the **number of statements** in the loop. In **horizontal** unrolling the expansion **appends to a single statement** .

The technique of unrolling is also used on non-vector, i.e. scalar machines. The performance improvement that one obtains on these machines however is not always transportable to vector machines, as will be shown below. In fact, on vector machines unrolling sometimes leads to performance degradation.

One should therefore be careful when applying this technique on vector computers.

Example 1 : Vertical Loop Unrolling

```
DO 10 I = 1,N
  DO 10 J = 1,4
    A(I,J) = B(I,J)*C(I,J)
10 CONTINUE
```

The inner loop has a small length (4) and can be expanded out inside the outer loop as follows :

```
DO 10 I = 1,N
  A(I,1) = B(I,1)*C(I,1)
  A(I,2) = B(I,2)*C(I,2)
  A(I,3) = B(I,3)*C(I,3)
  A(I,4) = B(I,4)*C(I,4)
10 CONTINUE
```

In this example unrolling will do a good job : a short vectorizable loop was replaced by a long vectorizable loop.

Example 2 :

```
DO 10 J = 1,N
  DO 10 I = 1,2*L
    A(I,J) = A(I,J) - B(I,J)
10 CONTINUE
```

Vertical loop unrolling can now be used to expand some of the iterations of the inner loop, e.g. :

```
DO 10 J = 1,N
  DO 10 I = 1,2*L,2
    A(I,J) = A(I,J) - B(I,J)                    ! s1
    A(I+1,J) = A(I+1,J) - B(I+1,J)              ! s2
10 CONTINUE
```

This technique, often used on scalar machines in order to reduce the loop overhead and to make better use of the cache memory, does not necessarily pay off on vector machines !

In this case, the unrolling does not lead to a longer loop, since the innermost loop is not eliminated anymore.

On a CRAY X-MP the above loop performs better in its original form.

In horizontal loop unrolling, iterations of a loop are appended onto a single statement.

Example 1 : Horizontal Loop Unrolling

```
        DO 10 J = 1,N
          DO 10 I = 1,3
            A(J) = A(J) - B(I,J)
     10 CONTINUE
```

The inner loop above is very short. If all its iterations are expanded, one obtains :

```
        DO 10 J = 1,N
          A(J) = A(J) - B(1,J) - B(2,J) - B(3,J)
     10 CONTINUE
```

Clearly, this is a horizontal loop unrolling.

If one compares both alternatives, the second one seems more advantageous both from the viewpoint of vector length (N, assumed large, compared to 3) as from the viewpoint of memory contention.

The original loop uses 3 memory references for 1 flop, whereas the modified version only needs 5 memory references for 3 flops.

Horizontal loop unrolling can also be applied to the outer loop as illustrated in the example below :

Example 2 :

```
        DO 10 J = 1,4M
          DO 10 I = 1,N
            Y(I) = Y(I) + X(J)*Z(I,J)
     10 CONTINUE
```

Assuming that M is large, an alternative, using horizontal outer loop unrolling, is :

```
        DO 10 J = 1,4M,4
          DO 10 I = 1,N
            Y(I) = Y(I) + X(J)*Z(I,J)       + X(J+1)*Z(I,J+1)
    +
                          X(J+2)*Z(I,J+2) + X(J+3)*Z(I,J+3)
     10 CONTINUE
```

The original loop requires 3 memory **vector** references for 2 flops (inside the inner loop X is a scalar). In the modified version, 6 memory vector references are needed to perform 8 flops (again X is a scalar inside the inner loop).

1.2.7.2 Loop Jamming **Loop jamming** is the process of combining different vectorizing loops into a single loop.

Loop jamming can be applied if two or more independent vectorizing loops have the same vector length. This technique allows to **eliminate** the **overhead of loop control** code of the eliminated loops.

Example :

```
      DO 10 I = 1,N
         A(I) = C(I)*D(I)
   10 CONTINUE
      DO 20 I = 1,N
         B(I) = C(I)/D(I)
   20 CONTINUE
```

The loops above have the same vector length and are both vectorizable. Hence it is advantageous to jam them into a single loop :

```
      DO 10 I = 1,N
         A(I) = C(I)*D(I)
         B(I) = C(I)/D(I)
   10 CONTINUE
```

1.2.7.3 Inverting Loops If in a nested loop the innermost loop has a smaller iteration count than one of the outer loops it **may be advantageous** to invert the loops.

However, one should be very cautious when applying this technique. In some cases it may cause performance degradation instead of performance improvement.

Some of the problems that may be caused by inverting loops are :

> • Appearance of bank conflicts.

> • Appearance of dependencies inhibiting the vectorization.

Example 1 :

```
      DO 10 I = 1,1000
         DO 10 J = 1,10
            A(I,J) = A(I,J) + B(I,J)*C(I,J)
   10 CONTINUE
```

The inner loop is much shorter than the outer loop. Inverting the loops leads to :

```
      DO 10 J = 1,10
         DO 10 I = 1,1000
            A(I,J) = A(I,J) + B(I,J)*C(I,J)
   10 CONTINUE
```

Here the loop inversion is appropriate : the same results as in the original version will be obtained but with a vectorizable inner loop that has a much larger iteration count than before. Note also that this inversion is favorable with respect to the occurrence of bank conflicts.

Example 2 :

```
DO 10 I = 1,1000
  DO 10 J = 1,10
    A(I,J) = A(I-1,J) + B(I,J)*C(I,J)
10 CONTINUE
```

This example is almost identical to the previous one; only the row index of A in the right-hand-side has been changed from I to I-1.

This minor change causes loop inversion to be inappropriate. Whereas the inner loop of the example above is still vectorizable, vectorization is inhibited if the loops are inverted, because of a recurrence on A.

1.2.7.4 Subprogram inlining   It was already mentioned before that subroutine calls within DO loops inhibit vectorization of that DO loop.
This **may** be overcome if the subroutine is **expanded inside** the loop.

Another advantage is that the **overhead** to transfer the control to the subroutine is **eliminated**. This overhead may be substantial : in many cases it takes up to 75 clock periods to call a subroutine with no arguments passed. With just one argument the overhead nearly doubles !

The same technique can be applied to external functions. They can be replaced by **statement functions** that do not inhibit vectorization of the loop from which they are called, as in the example below.

Example :

```
   ...
   DO 10 I = 1,N
     C(I) = FUN(A(I),B(I))
10 CONTINUE
   ...
   FUNCTION FUN(X,Y)
   FUN = (X+Y)**2
   END
```

The call of an external function in DO loop 10 inhibits the vectorization. If the external function is replaced by a statement function the loop vectorizes :

```
   FUN(X,Y) = (X+Y)**2
   ...
   DO 10 I = 1,N
     C(I) = FUN(A(I),B(I))
10 CONTINUE
   ...
```

Note that the statement function should appear in the program unit where it is used, and should be put before any executable statements.

1.2.7.5 Segmenting a non-vectorizing loop   In many cases a loop that can not be vectorized also contains **code that can be vectorized** and that is **not related** to the **non-vectorizable part**.
The technique of **loop segmenting** puts the vectorizable and the non-vectorizable part in **different loops**.

When applying this technique one should carefully check if the two parts to be separated are truly **unrelated**.

1.2.8. Optimization of basic vector operations

In this section some basic linear algebra constructs are considered and it is shown how these can be optimized. Some more details and more information about the use of supercomputers in linear algebra can be found in Gentsch (1984), Schönauer (1987) and van der Vorst (1988).

1.2.8.1. Matrix - vector multiplication   Consider a matrix-vector multiplication, as illustrated below

$$
\begin{pmatrix} Y_1 \\ Y_2 \\ \cdot \\ \cdot \\ Y_N \end{pmatrix} = \begin{pmatrix} A_{11} & A_{12} & \cdot & \cdot & A_{1N} \\ A_{21} & A_{22} & \cdot & \cdot & A_{2N} \\ \cdot & \cdot & \cdot & \cdot & \cdot \\ \cdot & & & & \cdot \\ A_{N1} & A_{N2} & \cdot & \cdot & A_{NN} \end{pmatrix} * \begin{pmatrix} X_1 \\ X_2 \\ \cdot \\ \cdot \\ X_N \end{pmatrix} \qquad (4.1)
$$

Component i of vector Y, Y_i, is calculated by multiplying row i of matrix A with the vector X, i.e.

$$
Y_i = \sum_{k=1}^{N} A_{ik} \cdot X_k \qquad (4.2)
$$

This is not a very efficient way of doing on most vectorcomputers since the elements of A are stored columnwise and hence have to be loaded with a large stride.

Another way to calculate this matrix-vector multiply is illustrated by equation (4.3)

$$
\begin{pmatrix} Y_1 \\ Y_2 \\ . \\ . \\ Y_N \end{pmatrix} = \begin{pmatrix} Y_1 \\ Y_2 \\ . \\ . \\ Y_N \end{pmatrix} + \sum_{j=1}^{N} X_j * \begin{pmatrix} A_{1j} \\ A_{2j} \\ . \\ . \\ A_{Nj} \end{pmatrix} \tag{4.3}
$$

In this way the elements of A are accessed column by column, i.e. in a contiguous way. The operation performed in (4.3) is the SAXPY operation.

If one looks at the Fortran implementation of (4.2) and (4.3) the only difference is an interchange in the loop indices.

Equation (4.2) would be coded as

```
      DO 10 I = 1,N
      DO 10 J = 1,N
         Y(I) = Y(I) + X(J)*A(I,J)
   10 CONTINUE
```

whereas equation (4.3) is coded as follows

```
      DO 10 J = 1,N
      DO 10 I = 1,N
         Y(I) = Y(I) + X(J)*A(I,J)
   10 CONTINUE
```

1.2.8.2. Matrix - matrix multiplication.   The Fortran coding of a matrix - matrix multiplication involves 3 levels of DO loops.
In the inner loop the arithmetic operations take place. By interchanging the loop indices six variants are possible.

Each of these variants has a different memory access pattern which will have an important impact on its performance on a vector processor.

The 'classical' way to code the multiplication C = A * B is

```
      DO 10 I = 1,N
      DO 10 J = 1,N
      DO 10 K = 1,N
         C(I,J) = C(I,J) + A(I,K)*B(K,J)
   10 CONTINUE
```

This can be described graphically using a diagram, as introduced by Dongarra (1983)

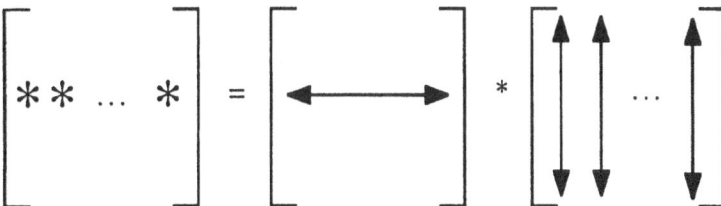

If the indices I and J are permuted the diagram reads

Both variants above perform an inner product in the most inner loop. The inner loop is therefore a reduction loop. Though such a loop is vectorizable, it is not very efficient. Moreover the elements of A are scanned column by column (2 succeeding elements are in different columns) which is not favorable for reasons of bank conflicts.

The algorithms with the indices in the order KIJ or KJI are related in that the inner loop is a SAXPY operation.

For the order KIJ, a row of B is multiplied with an element of A and the result is used to update a row of C. This option again manipulates matrix rows and therefore is not preferable.

The KJI alternative scales a column of A with an element of B to update a column of C. This option does not suffer from bank conflicts and is therefore the most efficient so far. However it should be noticed that, as the J index varies, another column of C has to be loaded and stored.

For each pass through the outer loop the complete matrix has to be loaded from the memory and stored again.

The total number of loads and storages can easily be evaluated.

For the two innermost loops (indices J and I) :

      • 1 load of column K of A

      • N loads of column J of C

      • N storages of column J of C

This whole process has to be repeated for the different indices of the outermost loop. This brings the total number of loads and storages, denoted T, to :

$$T = N.(2N + 1) = 2N^2 + N \tag{4.4}$$

The two final possibilities are IKJ and JKI.

The access pattern for the first option is

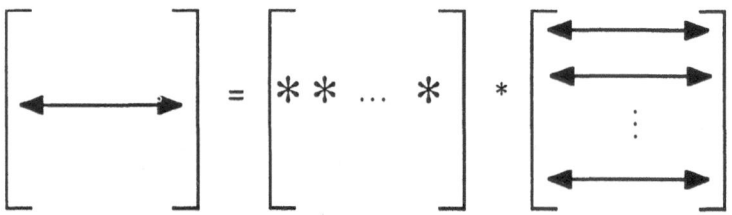

Here the inner loop is again a SAXPY loop but, as visualized above, the elements of B are accessed column by column, which may cause bank conflicts.

The graphical representation of the second option is :

This option is most appropriate on vector machines.

The Fortran implementation is given by

```
      DO 10 J = 1,N
      DO 10 K = 1,N
      DO 10 I = 1,N
         C(I,J) = C(I,J) + A(I,K)*B(K,J)
   10 CONTINUE
```

The inner loop consists again of a SAXPY operation, acting on contiguously stored vectors.

This was also the case with the KJI ordering, but, the present option requires less loads and storages.

If one considers the two innermost loops, one finds :

> • N loads of a column of A.

> • 1 load of column J of C

• 1 storage of column J of C

The count for the whole loop is therefore :

$$T = N.(N + 2) = N^2 + 2N \qquad (4.5)$$

Comparing equations (4.4) and (4.5) it turns out that the JKI ordering is the most favorable.

1.3 PARALLEL COMPUTERS

1.3.1. Introduction

When it comes to parallelization and multiprocessor machines, a first important distinction one has to make is between the *shared* and *distributed memory* systems.

In the first category all the processors share the same memory. As a result the number of processors cannot be unlimited since this would lead to memory contention problems. In practice, the number of processors can go up to 16 or 32 but not much more. Examples of such machines are the multiprocessor CRAY X-MP and Y-MP systems, the Silicon Graphics Power Challenge system. These are all MIMD (Multiple Instruction/ Multiple Data) type machines : the different processors can handle both different data and different instructions.

The shared memory systems offer an important advantage : applications can be parallelized automatically by the compiler. Since current compilers can not analyze the code beyond the subroutine level, the resulting parallelization will be *fine grained* (i.e. the tasks executed in parallel will be small). Typically the compiler will parallelize the DO loops of the code.

On multiprocessor CRAY computers, this automatic parallelization is called *autotasking*. The code is then analyzed by a preprocessor that inserts *compiler directives* in the code.

These compiler directives can also be inserted by the user itself. This is called *microtasking*. This is still a fine grained parallelism.

If one wants to exploit parallelism beyond the subroutine level, e.g. 2 subroutines executed in parallel, (i.e. *coarse grained* parallelism) the CRAY computers provide *macrotasking*. This type of parallelization can't be automated and the user has to insert the appropriate macrotasking directives inside his code.

However the use of microtasking and macrotasking has an important drawback : the parallelized code is not transportable to other non-CRAY computers.

A better alternative to macrotasking in our view is therefore the use of a shared memory version of a message passing system such as PVM (Parallel Virtual Machine). This will make the code transportable to other shared memory machines and also to distributed memory machines.

Coarse grained parallelization has the potential of higher speed-up than fine grained parallelization.

The distributed memory systems are theoretically not limited in number of processors. Existing monoprocessor applications however have to be adapted to run on these computers : the exchange of data (communication) between the different processors has to be included.

One distinguishes between SIMD and MIMD distributed memory systems.

In the SIMD paradigm all processors execute the same instruction but work on different data. This is also referred to as a *data-parallel* approach. SIMD computers, such as Connection Machine and MasPar, have a massive number of processors. In CFD applications one typically assigns one processor to each grid node, cf. Candler & Wright (1994), Blosch & Shyy (1994) for some recent applications.

In the MIMD model the processors can also handle different instructions. MIMD type computers are for instance the CRAY T3D, IBM SP1 and SP2, and clusters of workstations.

The parallelization on these machines is typically coarse grained : large tasks are executed in parallel with the communication (between processors) kept to a minimum. The latter is important as communication is an overhead that reduces the parallel efficiency.

CFD codes ported to MIMD architectures are usually based on a multiblock approach: different blocks of the mesh are attributed to different processors, Tysinger & Caughey (1993), Ryan & Weeratunga (1993), Drikakis & Schreck (1993), Lacor et al.(1994).

Message passing environments have been developed to make the communication machine-independent and, as a result, the codes more transportable. Different such environments are available , cf. McBryan (1994) for an overview, PVM (Parallel Virtual Machine) being one of the most popular.

In the remaining of this text we will only deal with coarse grained parallelization, using a message passing system.

As mentioned above, this makes the code transportable to both shared and distributed memory machines, and is therefore to be preferred to a machine dependent approach such as macrotasking.

On the other hand, the coarse grained parallelization offers a better potentiality for efficient parallelization than a fine grained parallelization approach, such as microtasking.

First some important concepts for parallelization will shortly be reviewed.

1.3.2. Some important concepts

1.3.2.1.Synchronization   **Synchronization** is a method of coordinating steps within tasks that run in parallel. A **synchronization point** is a point in time when a task receives a 'go-ahead signal' to proceed with its processing. That is, whatever the task was waiting for, has happened and a signal has been sent to and received by the waiting task.

This mechanism is shown schematically in figure 1. The horizontal axis represents time. A thick line indicates that the task is processing. The W indicates that the task is waiting for an event to occur. For instance, task 0 is waiting for an event and can only resume processing after task 1 posted the event occurrence (P on the task 1 line). Similarly, some time later task 1 has to wait for an event occurrence and can only resume after task 0 posted the event occurrence.

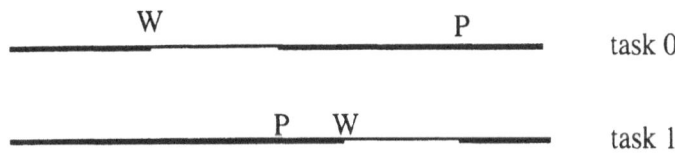

Figure 1 : Illustration of synchronization of two tasks.

It is clear that the synchronization increases the total run time of each task. Synchronization points should therefore be kept to a minimum. The possible delays should be taken into account in the **load balancing** technique, discussed hereafter.

1.3.2.2.Load balancing   A good load balancing ensures that each of the processors involved in a job, does approximately the same amount of work. All work that can be done in parallel is divided evenly among processors.

If all the work involved in a job can be done in parallel on n processors, and the load is balanced among them, the wall-clock time for the multitasked job can approach 1/n of the wall-clock time for the job run on one processor.

For a non well balanced code the speed up may be much less, because some of the processors will be idle most of the time. This is schematically illustrated in figure 2.

The horizontal axis again represents time. On top a code running on one processor is shown. Assume that the pieces 1 and 2 can be run in parallel as well as the pieces 3 and 4. Below the one processor line, a badly balanced 2 processor solution is shown. Processor P1 is idle most of the time, and only a small speed up is realized compared to the one processor case. The last solution is much better balanced, leading to a significant speed up.

C. LACOR

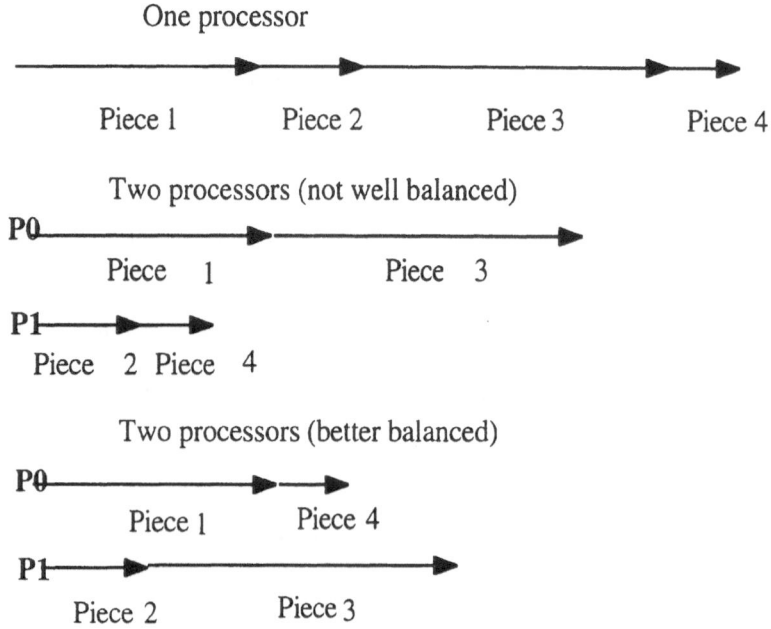

One processor

Piece 1   Piece 2   Piece 3   Piece 4

Two processors (not well balanced)

P0

Piece   1   Piece   3

P1

Piece   2 Piece   4

Two processors (better balanced)

P0

Piece 1   Piece 4

P1

Piece 2   Piece 3

Figure 2 : Load balancing on two processors.

A good load balancing is therefore very important for the parallel efficiency of the code.

There are two types of load balancing : **static** and **dynamic**.

In CFD applications with structured, fixed meshes one can use static load balancing : division of the mesh in blocks of similar size (which are then executed in parallel on different processors) will give a good load balancing.

In applications using unstructured meshes with adaptive refinement, the division of the mesh into blocks (and hence the load balancing) must necessarily be done dynamically, as the mesh changes through the iterative process. Tools are available to do this dynamic load balancing but this is beyond the scope of this lecture.

1.3.2.3.Critical regions   A critical region is a segment of code that accesses a resource, shared by different tasks which could be running concurrently. This resource can be Central Memory, I/O files, subroutines or anything else that is shared by the tasks in a job.
If for instance more than one task reads from and stores into shared memory locations, indeterminate results can arise. Neither task can be sure that the data it is reading is as expected, nor that the area of memory to which data is stored is ready to be overwritten.

Consider the following example :

```
      SUBROUTINE SUB(TOTAL,A)
      REAL A(5),TOTAL
      DO 10 I = 1,5
      . . .
       TOTAL = TOTAL + A(I)
       TOTAL = TOTAL * A(I)
      . . .
   10 CONTINUE
      RETURN
      END
```

The lines indicated in bold characters represent a critical region. Suppose that A(I) = FLOAT(I) and that the value of TOTAL passed in SUB is 0. The final outcome of this loop should then clearly be TOTAL = 645.

Suppose now that the iterations of this loop are executed concurrently and that no actions are taken to protect the critical region. For simplicity, assume that the first and the 2nd iteration are run in parallel. The outcome is then not predictable, it depends on the order in which the statements are executed. Two possibilities, leading to different results, are :

```
TOTAL = TOTAL + A(1)          TOTAL = TOTAL + A(2)
TOTAL = TOTAL * A(1)          TOTAL = TOTAL * A(2)
TOTAL = TOTAL + A(2)          TOTAL = TOTAL + A(1)
TOTAL = TOTAL * A(2)          TOTAL = TOTAL * A(1)
```

In the example to the left the statements are executed in the proper order, leading to the (correct) result TOTAL = 6. In the example to the right, the order of the statements is mixed up leading to an erroneous result TOTAL = 5.

From this example it may seem that the error comes from the fact that there is more than one instruction accessing shared data, and that the problem would be solved if one rewrites the loop as follows :

```
      SUBROUTINE SUB(TOTAL,A)
      REAL A(5),TOTAL
      DO 10 I = 1,5
      . . .
       TOTAL = (TOTAL + A(I)) * A(I)
      . . .
   10 CONTINUE
      RETURN
      END
```

However, the example above will not work either since one should keep in mind that a single instruction usually consists of different machine language instructions, of which the order can also be mixed up.

To ensure the correct processing in parallel mode of loops containing critical regions, locks are used. As soon as one process accesses the shared data it sets up a flag to indicate to the other processes that the shared variables are being used. At that moment no other process can enter the critical region. This of course assumes that each process checks the lock before entering a critical region. As soon as the process leaves the critical region it clears the lock.

Because a process which is unable to enter a critical region is forced to wait, it is important to keep the length of critical regions (in execution time) to a minimum. On the other hand, if one has many small critical regions, there is a high overhead due to the cost of the locking operations.

1.3.2.4. Amdahl's law   This law gives the speed up or the gain G attainable on a multiprocessor machine (compared to a one processor machine) as a function of the number of processors P and the percentage f of the code that can be run in parallel.
This law is very similar to the one formulated for vectorization.

Suppose T is the total execution time on 1 processor.
On the multiprocessor machine a percentage 1-f still has to be executed on 1 processor. The corresponding time is (1-f)*T.
The remaining f percent of the code is executed on the P processors. Neglecting synchronizations and assuming a perfect load balancing, the speedup for this part of the code will be P. It therefore can be executed in f*T/P seconds.

Hence, the gain G is

$$G = \frac{1}{(1-f) + \frac{f}{P}} \tag{1}$$

This formula is very similar to the one derived for the vectorization.

Since G is a function of two parameters, f and P, one can consider for simplicity the case of an infinite number of processors, i.e. $P \to \infty$. This gives the maximum performance gain if a percentage f of the code runs in parallel :

$$G_{max} = \frac{1}{1-f} \tag{2}$$

This is completely identical to the expression for the speed up of a code if a percentage f is vectorized, assuming the vector functional units are infinitely fast. The conclusions which can be drawn from this law are also analog :

  • A high percentage of the code has to run in parallel before a significant gain can be achieved.

  • An improvement in the software, so that a higher percentage of the code can run in parallel, pays off more than a hardware improvement, i.e. than having a machine with much more processors.

The law can be extended to take the overhead associated with multitasking into account. Assuming that this overhead takes α% of the total execution time on 1 processor, the expression for the gain becomes :

$$G = \frac{1}{(1\text{-}f) + \frac{f}{P} + \alpha} \tag{3}$$

1.3.3. Parallelization of CFD codes

As already mentioned in the introduction, a straightforward parallelization (on shared memory computers) is on the level of the compiler. This requires no effort from the user, and allows to obtain already reasonable speed-ups.

However coarse grained parallelization offers a better potential.

Here we will discuss only the parallelization using a portable message passing system, and no other approaches, such as CRAY macrotasking, cf. introduction, which is not transportable.

Also we will focus on MIMD computers. Programming SIMD computers is more complex. On the other hand MIMD computers are much more available; it suffices to have some workstations connected by a network.

Virtually all CFD applications on MIMD type computers use the *domain decomposition* technique for the parallelization. The calculation domain is artificially split into different *domains* or *blocks*.

In principle, different blocks are assigned to different processors. In practice, for complicated geometries, the mesh may already contain a number of blocks which is higher than the number of processors. In that case more than 1 block must be assigned to the processors. Also, if the mesh contains some smaller blocks, it is better, for reasons of load balancing, not to assign these small blocks to separate processors.

Good load balancing is achieved by assigning each processor approximately the same number of nodes. One can further refine this by distinguishing between inner nodes and boundary nodes (which require a different treatment and hence contribute differently to the load).

On the boundaries between different blocks information has to be exchanged. This is achieved by communication through message passing. This communication corresponds to an overhead and should be kept to a minimum. Referring to Amdahl's law, what is important is the fraction of the total work that is taken by communication. E.g. a turbulent Navier-Stokes calculation may have a good parallel efficiency, whereas the parallel efficiency of an Euler simulation on the same mesh will be much less (assuming the data to be communicated are the same).

Although one will certainly gain an important factor in wall clock execution time per iteration for a multi-block, multi-processor calculation compared to a single block, single processor calculation, one must also account for a degradation in convergence rate between the single and the multi-block case. The multi-block case might require more iterations to converge than the single block.

This is especially true for implicit schemes. E.g. for line relaxation in a single block, all the unknowns along the line, traversing the complete block, are coupled; in a multi-block simulations, the lines are restricted to the length of the blocks.

Also when using explicit schemes, such as Runge-Kutta time stepping, there is a difference in convergence rate between the single and multi-block case, if one does not treat the connections between blocks carefully. Though the convergence degradation usually is only minor, it also implies that, if the single block calculation is time accurate, the multi-block calculation is not time accurate anymore.

This will discussed into more detail in the next chapter on Parallel Algorithms.

2. Parallel algorithms

2.1. INTRODUCTION

In this chapter parallel algorithms for Computational Fluid Dynamics (CFD) applications will be discussed.

We restrict ourselves to the MIMD class of computers, distributed memory or shared memory.

On the distributed memory machines, message passing must be used to communicate between different processors.

As pointed out in section 1.3.1, adapted versions of the message passing libraries are available on most shared memory machines.

A code developed to run on a distributed memory machine is therefore also transportable to those shared memory machines.

For this reason we will restrict ourselves to the use of message passing libraries and not discuss other means for parallelization on shared memory machines, such as macrotasking on CRAY computers.

The most straightforward way to parallelize the CFD code is then to use a *domain decomposition* (also *multi-block*) approach.

2.2. SPATIAL DISCRETIZATION

The Euler/Navier-Stokes system in multi-dimensions can be expressed as

$$\frac{\partial U}{\partial t} + \vec{\nabla}.\vec{F} = 0 \qquad (1)$$

with U the conservative variables and $\vec{F}$ the flux vector with components f,g,h in resp. x,y,z direction.

Discretization in space with a control volume approach, based on a *cell-centered* method, leads to :

$$\int_{\Omega} \frac{\partial U}{\partial t} d\Omega + \sum_{\text{faces}} \left(\vec{F}.\vec{n}\right)^{*} \Delta S = 0 \qquad (2)$$

where Ω, ΔS, $\vec{n}$ are resp. cell volume, area of the cell face and cell face normal, and $\left(\vec{F}.\vec{n}\right)$ is the numerical flux through the cell face.

In a cell-centered method, the unknowns U are stored in the centers of the mesh cells. In a *cell-vertex* method, the unknowns are stored in the vertices of the mesh cells. The main difference between cell-centered and cell-vertex methods, lies in the treatment of the boundary conditions; in a cell-vertex method one stores the solution on the boundaries itself, which is not the case for a cell-centered method.

The multi-block strategy, discussed below, is very similar for both type of discretizations, and in the following it is tacitly assumed that a cell-centered approach is adopted.

Eq.(2) is valid both for *unstructured* and *structured* meshes. Again the multi-block strategy can be used for both types of meshes. With an unstructured mesh, the load balancing is more difficult and one has to use dynamic load balancing techniques, especially if grid adaptation (i.e. the dynamic adaptation of the grid during the calculations) is used. This however is beyond the scope of this lecture, and in the remainder we will only deal with structured meshes.

In structured meshes the cells can be identified by i,j and k indices. The numerical flux can then generally be expressed as

$$\left(\vec{F}.\vec{n}\right)^{*}_{i+1/2} = \frac{1}{2}\left\{\left(\vec{F}.\vec{n}\right)_{i} + \left(\vec{F}.\vec{n}\right)_{i+1}\right\} - d_{i+1/2}$$
$$(3)$$

where the indices j,k which do not vary, are dropped in equation (3) as will consistently be done in the remaining of this text in order to enhance clarity.

Expression (3) leads to either *central* or *upwind* schemes, depending on the dissipation term $d_{i+1/2}$ used in the right-hand-side.

The *central* scheme results from a Jameson type dissipation, Jameson et al.(1981), using a blending of 2nd and 4th order derivatives of the conservative variables

$$d_{i+1/2} = \varepsilon^{(2)}_{i+1/2}\, \delta U_{i+1/2} + \varepsilon^{(4)}_{i}\delta^{2}U_{i} - \varepsilon^{(4)}_{i+1}\, \delta^{2}U_{i+1} \qquad (4)$$

The scalar coefficients ε are given by :

$$\varepsilon^{(2)}_{i+1/2} = \kappa^{(2)}\, \lambda^{*}_{i+1/2}\mathrm{max}\left(v_{i-1}, v_{i}, v_{i+1}, v_{i+2}\right)$$

$$\varepsilon^{(4)}_{i+1/2} = \mathrm{max}\left(0, \kappa^{(4)}\lambda_{i+1/2} - \varepsilon^{(2)}_{i+1/2}\right)$$
$$(5)$$

The cell centered values of $\varepsilon^{(4)}$ in eq.(4) are obtained by arithmetic averaging of the cell face values of eq.(5). The variables v_i are sensors to activate the second-difference dissipation in regions of strong gradients, such as shocks, and to de-activate it elsewhere. They measure variations of pressure and are defined as :

$$v_i = \left| \frac{p_{i+1} - 2p_i + p_{i-1}}{p_{i+1} + 2p_i + p_{i-1}} \right| \tag{6}$$

$\lambda*$ in eq.(5) is a measure of the inviscid fluxes and is commonly chosen as the spectral radius multiplied with the cell face area :

$$\lambda^{*}_{i+1/2} = \left(\vec{v}.\vec{\Delta S} + c\Delta S \right)_{i+1/2} \tag{7}$$

In the *upwind* case, the dissipation term is more complicated. The numerical dissipation is now related to the variation of the conservative variables through a matrix instead of scalar coefficients :

$$d_{i+1/2} = \frac{1}{2} R_{i+1/2} \, \mathrm{diag}\!\left(\alpha^l_{i+1/2}\right) R^{-1}_{i+1/2} \, \delta U_{i+1/2} \tag{8}$$

The matrices R and R^{-1} contain resp. the right and left eigenvectors of the Jacobian matrix. $\mathrm{diag}\!\left(\alpha^l_{i+1/2}\right)$ in eq.(8) represents a diagonal matrix with $\alpha^l_{i+1/2}$ the element in row and column l.

Different upwind schemes result, depending on the expression for α.

For a TVD version of the Flux Difference Splitting scheme (FDS TVD), Roe (1981), and Symmetric TVD(STVD), Yee (1987), α can be expressed as :

$$\alpha_{i+1/2} = \left| \lambda_{i+1/2} \right| \left[1 - Q_{i+1/2} \right] \tag{9}$$

where $\lambda_{i+1/2}$ represents the eigenvalues of the Jacobian matrix and $Q_{i+1/2}$ is a limiter function, ensuring monotonicity.

The limiter $Q_{i+1/2}$ acts on ratios $r_{i+1/2}$ of variations of the characteristic variables, defined as :

$$r^{-}_{i+1/2} = \frac{w_i - w_{i-1}}{w_{i+1} - w_i} , \; r^{+}_{i+1/2} = \frac{w_{i+2} - w_{i+1}}{w_{i+1} - w_i} \tag{10}$$

For the FDS TVD scheme, Q is function of only one of these ratios; r^- if $\lambda_{i+1/2}$ is positive and r^+ if $\lambda_{i+1/2}$ is negative, i.e. information into the direction where the signal comes from, is used. The 'classical' limiters such as minmod, van Leer, van Albada, superbee,.. can be used as they are all defined as functions of one ratio.

In the STVD scheme, the check on the eigenvalue sign is avoided by choosing Q as a function of the two ratios r^- and r^+ :

$$Q = Q\!\left(r^{-}_{i+1/2}, r^{+}_{i+1/2} \right) \tag{11}$$

Eq. (11) excludes the limiters mentioned above, as Q is now function of 2 ratios instead of one. To circumvent this problem, Davis (1984), in the framework of Lax-Wendroff TVD schemes, suggests to use both + and − limiters :

$$Q = Q\!\left(r_{i+1/2}^-\right) + Q\!\left(r_{i+1/2}^+\right) - 1 \tag{12}$$

A novel approach is choosing the limiter function Q as, Lacor et al.(1993) :

$$Q = Q\!\left(r^*\right) \tag{13}$$

with r* a so-called effective ratio, obtained by averaging r+ and r-. It can be shown that any of the limiters mentioned above, when applied to r* leads to a monotonic scheme provided $r^* \geq 0$, Lacor et al.(1993).

Some possible definitions of r* are:

$$\min\!\left(r^+,r^-\right) \leq \frac{r^+\!\left(r^-\right)^2 + r^-\!\left(r^+\right)^2}{\left(r^+\right)^2 + \left(r^-\right)^2} \leq \frac{2r^+r^-}{r^+ + r^-} \leq$$
$$\sqrt{r^+r^-} \leq .5\!\left(r^+ + r^-\right) \leq \max\!\left(r^+,r^-\right) \tag{14}$$

It can also be shown that definitions with higher values lead to a more compressive limiter and hence a less dissipative scheme.

2.3. SOLVERS AND MULTIGRID

The multigrid method is a very efficient tool to accelerate the convergence. It can be combined both with explicit and implicit solvers, which act as 'smoothers', i.e. they damp out the high frequency errors.

For non-linear equations, such as the Navier-Stokes equations, the multigrid Full Approximation Scheme is most suited. Since this strategy is quite standard, it is only shortly described below.

Consider a set of meshes denoted with an index $l = 1,...,L$ with L the finest level. The Navier-Stokes problem on the finest level can be written as :

$$\frac{\partial U^L}{\partial t} + N_L(U^L) = 0 \tag{15}$$

where $N_L(U^L)$ is the spatial discretization of the Navier-Stokes operator on the finest mesh L. Note that the temporal discretization has been left unspecified so far.

The problem is then approximated on coarser levels l as :

$$\frac{\partial U^l}{\partial t} + N_l(U^l) = F_l \tag{16}$$

with F_l the forcing function, defined recursively as :

$$F_l = N_l \left(I_{l+1}^l \; U^{l+1}\right) + \hat{I}_{l+1}^l \left[F_{l+1} - N_{l+1}\left(U^{l+1}\right)\right] \tag{17}$$

where I_{l+1}^l and $\hat{I}_{l+1}^l$ represent restriction operators of resp. the unknowns and the residuals.

After temporal discretization, eq.(16) becomes :

$$S.\Delta U^l + N_l (U^{l\,(0)}) = F_l \qquad (18)$$

$U^{l\,(0)}$ is the current solution on mesh l, around which the equations have been linearized (in an implicit method) and which has to be smoothed. One has:

$$U^{l\,(0)} = I^l_{l+1}\, U^{l+1} \qquad (19)$$

ΔU^l is an update of $U^{l(0)}$ and is to be calculated.

S is the smoother. It is the operator that corresponds to the chosen time integration method.

If the time integration is explicit, S is a diagonal matrix. An efficient explicit solver is Runge-Kutta time stepping, which, for steady state calculations, can be combined with local time stepping and with residual smoothing to enhance the efficiency. By carefully choosing the Runge-Kutta coefficients, it is also a good smoother for the multigrid.

If the time integration is implicit, the operator S is much more complicated, and eq.(18) corresponds to a linear system coupling all the unknowns. In relaxation type methods, the system (18) is then solved iteratively with a point or line relaxation method. In general, one or few relaxation sweeps suffice to smooth the solution.

Once the linear problem (18) is solved, the updated solution U^l (which will be smooth provided S is a good smoother) can be restricted to the next coarser level, according to eq. (19) with l replaced by l-1.

Once the solution on the coarsest mesh is smoothed, the coarse-to-fine sweep of the multigrid cycle is initiated. The current solutions on finer grids are updated with the solution on the next coarser level :

$$U^l = U^l + I^l_{l-1}\, (U^{l-1} - I^{l-1}_l\, U^l) \qquad (20)$$

The operator I^l_{l-1} is a prolongation operator.

The new solution on the finer mesh can be smoothed before proceeding to the next finer level, either by solving eq. (18), with $U^{l(0)} = U^l$, or by applying the residual smoothing operator to the corrections.

The computing cost per multigrid cycle can significantly be reduced on coarser grids with some simplifications, such as the use of first-order accurate schemes. This does not destroy the higher accuracy on the finest mesh, and also helps in improving robustness.

2.4 MULTIBLOCK IMPLEMENTATION

In order to mesh complex geometries with a structured approach, a multiblock strategy is almost unavoidable. At the same time, this multiblock strategy can be used for the parallelization.

Many multiblock codes can specify only one boundary condition per block face. While this simplifies the boundary treatment, it leads to a large number of relatively small-size blocks, often numbering in the hundreds for even rather simple configurations. For load balancing on parallel computers and overall efficiency on vector computers, such large number of small blocks is to be avoided. This can be done by allowing any possible number of boundary conditions on any block face.

On the other hand, the code should not impose any restriction on the number of blocks, allowing any block to be artificially split up into more blocks, if for instance the memory of the available processors or the number of processors requires so.

In the EURANUS code, developed by VUB and FFA, so-called ghost cells are used to store, for each block, the information of connected blocks. Fig. 2.4.1 shows 2 blocks that share a common interface. Each block has 2 rows of ghost (or dummy) cells, indicated by the dashed lines in fig. 2.4.1.

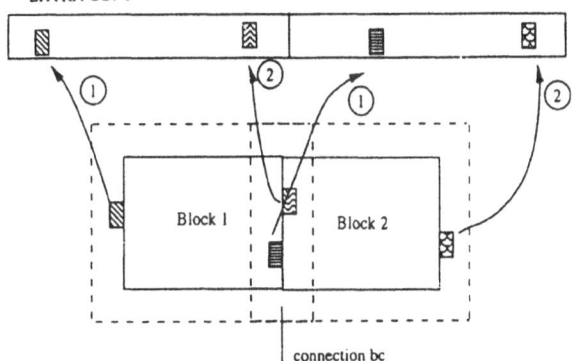

Figure 2.4.1. Use of extra dummy cells for multiblock on a single processor with all data in memory

The *ghost* cells of block 1 at the interface are used to store the data in the 2 rows of *inner* cells of block 2 at the interface, and vice versa. Two rows are needed for second-order accurate schemes.

On the non-connecting block sides, the first (i.e. immediately adjacent to the side) dummy cell row is used to satisfy the boundary conditions. The second row is superfluous here but is kept to facilitate the programming.

These arrangements suffice to run in multiblock. However the convergence for the monoblock and multiblock case will be different. This is easy to see. Take the 2 block case of fig. 2.4.1 as example. After updating the solution in the inner cells of block 1, the dummy cells have to be updated. On non-connecting boundaries the first dummy cell row of block 1 is updated. On the connecting boundary however, the new values in the 2 rows of inner cells of block 1 near the boundary have to be put in the dummy cells of *block 2*. As a result, when calculating the fluxes in block 2, these will differ from the fluxes of the monoblock case near the connection boundary, leading to different results.

Similarly, the multiblock convergence also depends on the order in which blocks are treated.

Though the multiblock calculation will not necessarily be slower that the monoblock, it is surely an unwanted effect for time accurate solutions, where all blocks should be treated in phase.

This problem can be solved by keeping an extra copy of the dummy cells in memory, cf. fig. 2.4.1.

After updating the inner cells of block 1, the new dummy cell values are now deposited in the extra copy (arrows 1, fig. 2.4.1). Note that the values of the inner cells of block 1 near the connection are stored in the extra copy of dummy cells of block 2. Next the other blocks are treated in the same way (arrows 2, fig.2.4.1). After treating all blocks, the extra copy of dummy cells contains the updated information and is copied to the physical dummy cells, fig.2.4.2.

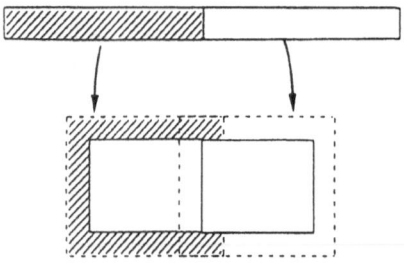

Figure 2.4.2. Extra dummy cells are copied into physical dummy cells for multiblock on a single processor with all data in memory

For small memory computers, one can also introduce the possibility to process one block at a time, write the results to disk, proceed to the next block, and so forth. In order to treat the blocks in phase, 2 extra copies of dummy cells are then needed, as illustrated in fig. 2.4.3.

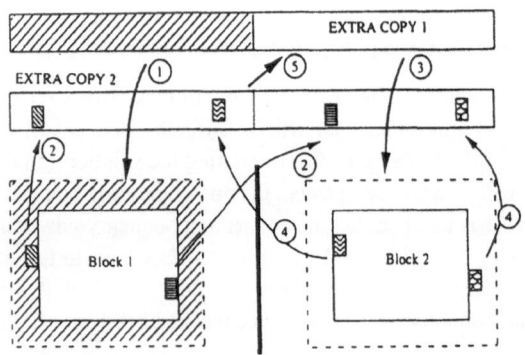

Figure 2.4.3. Use of extra dummy cells for multiblock on a single processor with blocks one by one into memory

As one block at the time is in memory, only part of the first extra copy (containing updated boundary information) can be used (arrow 1). Next the inner cells of block 1 are updated, after which the updated boundary information is stored in the 2nd extra copy of dummy cells (arrow 2). Note that if only one copy would be available the boundary information for block 2 would be modified. With 2 extra copies this is avoided and the boundary information for block 2 can be copied into the dummy cells of block 2, as soon as this block is in memory (arrow 3). Updating of block 2 leads to a new modification of the 2nd extra copy of dummy cells (arrow 4). After treating all blocks this extra copy contains the updated boundary information. It is now completely copied into the first extra copy (arrow 5), before restarting the procedure described above.

On a distributed memory system, there is not a shared memory to put the extra copy of dummy cells. Each block has now its own copy, fig. 2.4.4. The procedure is similar as for the monoprocessor case with all data in memory. The difference is that each block writes updated boundary information in its own extra copy if it concerns a non-connected boundary and in the extra copy of the connected block if it is a connected boundary, fig.2.4.4. The latter operation requires a communication between the connected blocks, indicated with thicker arrows in fig. 2.4.4.

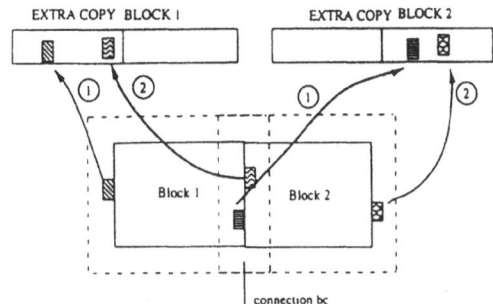

Figure 2.4.4. Use of extra dummy cells for multiblock on multiprocessor.

Note that in the current implementation a complete (i.e. for all blocks) set of extra dummy cells is stored on each processor. This is not really needed but it reduces the adaptation from the serial to the parallel version to a minimum.

Load balancing is ensured by attributing approximately the same number of points to the different processors.

In the EURANUS code the Fortran grid loop is outside the block loop, cf. fig. 2.4.5.

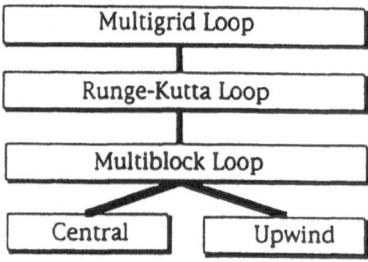

Figure 2.4.5. Order of Fortran loops in the EURANUS code

This means that any grid level is treated in *phase on all blocks*, referred to as horizontal mode by Tysinger & Caughey (1993).

Combined with the Runge-Kutta solver and the treatment of boundary conditions described above, this approach ensures that convergence (in terms of number of cycles) is exactly the same for a mono- and a multiblock case. This also means that the blocks communicate on every grid level and at each stage. Though this is needed for time-accuracy, this may be relaxed for steady state simulations.

Alternatively, the block loop can be put outside the grid loop, so-called vertical mode, Tysinger & Caughey (1993). This requires less communication between blocks but the previous property is not ensured anymore.

2.5 SOME RESULTS

As illustration, the Euler flow around a 2D NACA0012 airfoil at M=0.8 and 1.25 degrees angle of attack is chosen. From the viewpoint of parallel speed-ups, a 2D Euler simulation is more critical than a 3D Navier-Stokes simulation, as the ratio of computational work versus communication is lower, leading to a more pronounced influence of the communication on the speed-up.

Results are obtained on 5 different platforms of MIMD type : an IBM SP1 computer, a Parsytec GC-1/32 computer (with 32 T805 transputers), and 3 different clusters of workstations; a cluster of IBM RS6000 at the Université Libre de Bruxelles (ULB), another cluster with HP735 and DECα machines at the Vrije Universiteit Brussel (VUB) and a cluster with SGI R3000 and SGI Challenge at FFA, Sweden.

On all platforms, except on Parsytec, PVM was used for the communication. On the Parsytec machine the communication routines provided by the PARIX Operating System were used.

Three meshes were used : a fine (257*65), medium (129*33) and coarse (65*17) mesh. The latter one was only used on the Parsytec, where the finer meshes did not fit into the relatively small memory.

Figure 2.5.1 shows the convergence in terms of multigrid cycles for the fine and medium mesh. Four grid levels were used and a W-cycle. Five stage Runge-Kutta is used as smoother, and combined with implicit residual smoothing. Convergence to machine accuracy is achieved in less than 90 cycles for both meshes. As mentioned above, the same convergence curve is obtained irrespective of the number of blocks used.

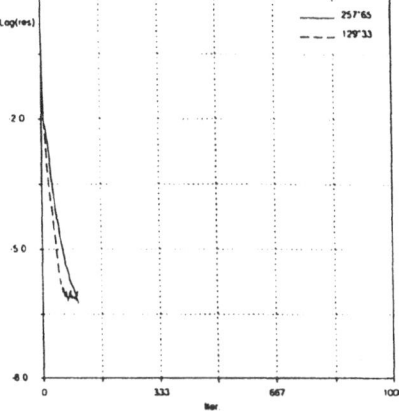

Figure 2.5.1 : Convergence history for NACA0012 (M=0.8, α=1.25°) on 257*65 and 129*33 mesh.

The aim is not to compare CPU timings between the different platforms but rather to assess the possible speed up on each platform. This speed up is defined as the CPU time per iteration on 1 processor divided by the CPU time per iteration on multiprocessors, on the same (multiblock) mesh. By also using a multiblock mesh on 1 processor (instead of single block) the difference between measured and theoretical speed up is a measure of the influence of the communication only.

In a first series of tests parallel calculations on the medium mesh were done. Multiblock meshes were generated with blocks connected in 'a serial way', as shown in figure 2.5.2 for the 8 block case.

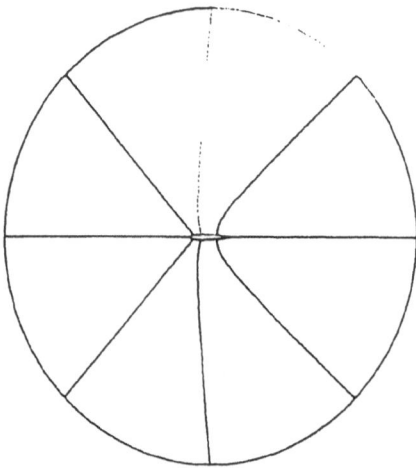

Figure 2.5.2 : 8 block mesh for NACA0012.

The obtained speed ups are depicted in figures 2.5.3, 2.5.4

Figure 2.5.3 shows results for the ULB cluster (8 IBM RS6000) and the VUB cluster (2HP735 and DECα). Both clusters consist of ethernet networks. The communication on the ethernet network is rather slow, which is clearly visible with speed ups rather far away from the theoretical maximum, even for this small number of processors. It should be added that the 6 blocks mesh consisted of 4 blocks with 20 cells in circumferential direction and 2 blocks with 24 cells, causing some load unbalancing. In addition, because of the 20 cell blocks only 3 level multigrid could be used there.

The VUB speed up is better than for ULB; the reason is probably that on the VUB cluster, although PVM started up resp. 3 and 5 processes (2 and 4 nodes and one host process), only 2 and 3 machines were used. Hence more than 1 process was running on the same machine. Though this will slow down this machine, this effect seems largely compensated by the improved communication speed between these 2 processes.

Figure 2.5.4 shows results for the SGI cluster, SP1 and Parsytec. On the SGI cluster good speed ups were obtained, comparable to those on the Parsytec. The results on the latter machine were obtained on the coarse 65*17 mesh, the medium mesh being too big to run on 1 processor.

Though SP1 gives better results than the VUB and the ULB cluster, the speed up is somewhat lower than for SGI and Parsytec.

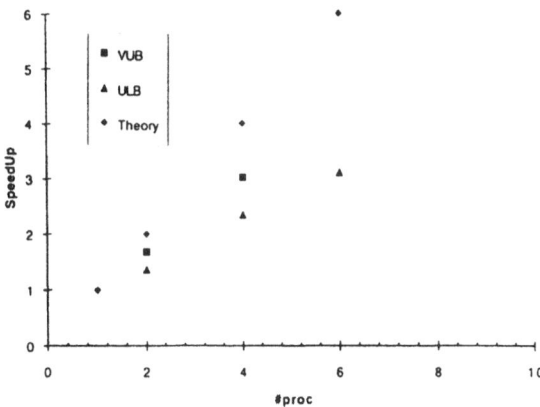

Figure 2.5.3 : Speed Up on 129*33 mesh for VUB, ULB cluster.

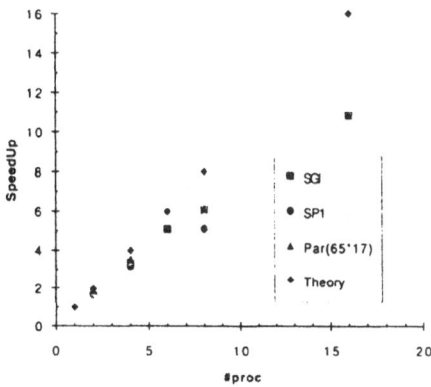

Figure 2.5.4 : Speed Up on 129*33 mesh for SGI cluster, SP1 and Parsytec

Next the same tests were repeated on the fine 257*65 mesh, figures 2.5.5 to 2.5.7

It is seen that on all platforms an improvement in speed up is obtained : as more computational work is to be done the relative importance of communication goes down.

For the VUB, ULB clusters some problems with PVM were noted, which sometimes got stuck, and which explains the few data points only available.

On the Parsytec the 129*33 mesh could be run on 16 processors but not on one. The CPU time on 1 processor (needed to find the speed up) was then estimated by

C. LACOR

extrapolating the 1 processor CPU time on the 65*17 mesh, using the ratio of 1 processor HP735 CPU timings on both grids.

For SP1 the speed up on the fine mesh with 8 processors seems somewhat in contradiction with the other SP1 speed ups. The reason is not clear, but it may again be due to the use of wall clock times in the calculation of speed up.

All calculations above used multigrid with communication between the processors on each grid level and in each Runge-Kutta stage .

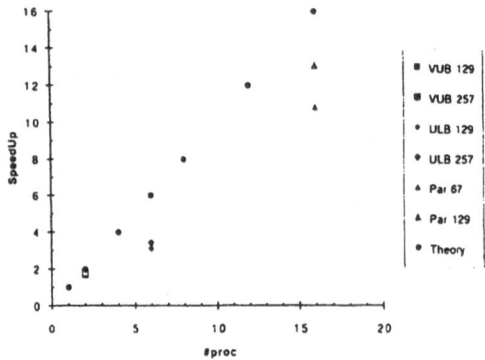

Figure 2.5.5 : Comparison Speed Up on 257*65 and 129*33 mesh for VUB, ULB cluster and Parsytec.

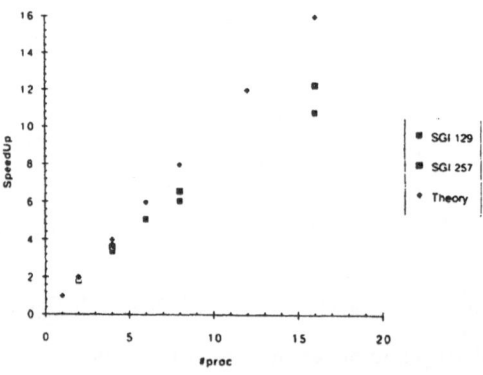

Figure 2.5.6 : Comparison Speed Up on 257*65 and 129*33 mesh for SGI cluster.

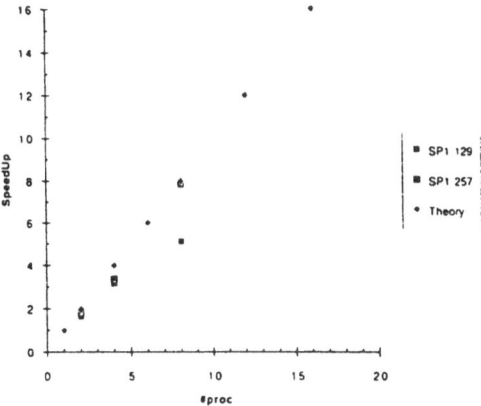

Figure 2.5.7 : Comparison Speed Up on 257*65 and 129*33 mesh for SP1.

In order to evaluate the relative importance of the communication on the coarser levels, calculations on the medium mesh were repeated in single grid on one platform. The ULB cluster was chosen as here the effect of communication is more pronounced. Fig. 2.5.8 shows the results. No major improvement in speed up is seen when running single grid. The conclusion is that the coarser grid communication takes much less time than communication on the finest level, where of coarse the number of data to be exchanged is larger. This result is in contrast with results of Tysinger & Caughey (1993) who find (with an implicit solver) a major improvement in speed up when switching from single to multigrid.

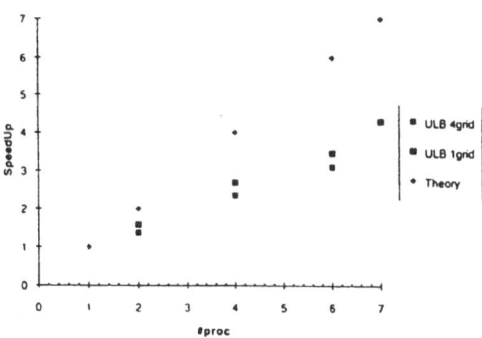

Figure 2.5.8 : Comparison Speed Up single- and multigrid on 129*33 mesh for ULB cluster.

Finally, the influence of different block arrangements on the speed up was investigated. The 129*33 mesh was tested on the SGI cluster using the following block arrangements

: 4*1 and 2*2 (4 blocks) and 16*1, 4*4 and 8*2 (16 blocks). Results are depicted in fig. 2.5.9. Although differences are small, the best strategy seems to keep the number of communication boundaries to a minimum (4*1 and 16*1).

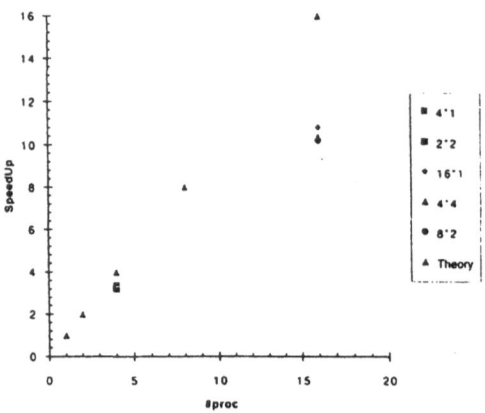

Figure 2.5.9 : Comparison Speed Up for different block configurations on 129*33 mesh for SGI cluster.

References

Blosch E.L., Shyy W. (1994) 'Parallel Efficiency of Sequential Pressure-Based Navier-Stokes Algorithms on the CM-2 and MP-1 SIMD Computers', AIAA paper 94-0409, AIAA 32nd Aerospace Sciences Meeting & Exhibit, Reno,.

Candler G.V., Wright M.J. (1994) 'A Data-Parallel LU-SGS Method fro Reacting Flows', AIAA paper 94-0410, AIAA 32nd Aerospace Sciences Meeting & Exhibit, Reno,.

Davis S.F. (1984) 'TVD Finite Difference Schemes and Artificial Viscosity', ICASE Report 84-20, NASA CR 172373, NASA Langley Research Center.

Dongarra J.J. (1983). ' Redesigning linear algebra algorithms', Proc. 1st International Coll. on Vector and Parallel Computing in Scientific Appl., Bulletin de la Direction des Etudes et Recherches, Serie C, pp. 51-59.

Drikakis D., Schreck E. (1993) 'Development of Parallel Implicit Navier-Stokes Solvers on MIMD Multi-Processor Systems', AIAA 31st Aerospace Sciences Meeting & Exhibit, Reno.

Gentsch W. (1984). 'Vectorization of Computer Programs with Applications to Computational Fluid Dynamics', Notes on Numerical Fluid Mechanics, Volume 8, Vieweg.

Gentsch W., Neves K.W. (1988).'Computational Fluid Dynamics : Algorithms and Supercomputers', AGARDograph No. 311 (Ed. H. Yoshihara).

Hockney R.W., Jesshope C.R. (1988). 'Parallel Computers 2', Adam Hilger, Bristol.

Jameson A., Schmidt W., Turkel E. (1981) 'Numerical Solutions of the Euler Equations by Finite Volume Methods using Runge-Kutta Time-Stepping Schemes', AIAA-81-1259.

Lacor C., Hirsch Ch., Eliasson P., Lindblad I. (1994) 'Study of the Efficiency of a Parallelized Multigrid/Multiblock Navier-Stokes Solver on Different MIMD Platforms', 2nd European CFD Conf., Stuttgart, 5-8 September.

Lacor C., Zhu Z.W., Hirsch Ch. (1993) 'A new Family of Limiters Within the Multigrid/Multiblock Navier-Stokes Code EURANUS', AIAA/DGLR 5th Int. Aerospace Planes and Hypersonics Technologies Conf., Munich.

McBryan O.A. (1994) 'An overview of message passing environments', Parallel Computing, Vol. 20, pp. 417-444.

Roe P.L. (1981) 'Approximate Riemann solvers, parameter vectors and difference schemes', J. Comp. Physics, Vol. 43, pp. 357-382.

Ryan J.S., Weeratunga S.K. (1993) 'Parallel Computation of 3-D Navier-Stokes Flowfields for Supersonic Vehicles', AIAA paper 93-0064, AIAA 31st Aerospace Sciences Meeting & Exhibit, Reno.

Schönauer W. (1987). 'Scientific Computing on Vector Computers', Special Topics in Supercomputing, Volume 2, North-Holland.

Tysinger T.L., Caughey D.A. (1993) 'Distributed Parallel Processing Applied to an Implicit Multigrid Euler /Navier-Stokes Algorithm, AIAA paper 93-0057, AIAA 31st Aerospace Sciences Meeting & Exhibit, Reno.

Van der Vorst H. (1988). 'Parallel Rekenen en Supercomputers', Academic Service.

Yee H.C. (1987) 'Construction of Explicit and Implicit Symmetric TVD Schemes and their Applications', Journal of Computational Physics, Vol. 68, pp. 151-179.

PORTING INDUSTRIAL CODES ON HIGH-PERFORMANCE COMPUTERS

and distributed computing using PVM

PATRICK R. AMESTOY AND MICHEL J. DAYDÉ

ENSEEIHT-IRIT, 2 rue Camichel, Toulouse, FRANCE

Abstract.

We address the main issues when porting existing codes from serial to parallel computers and when developing portable parallel software on MIMD multiprocessors (shared memory, virtual shared memory, distributed memory multiprocessors, and networks of computers). We especially address distributed/heterogeneous computing on multiprocessors and networks of computers using the PVM programming environment.

We illustrate this by using examples from our experience in porting industrial codes and in designing parallel numerical libraries. We report in some detail on the optimization and the parallelization of scientific applications coming from Centre National d'Etudes Spatiales, from Aérospatiale, and from the ESPRIT III EUROPORT-1 'PARALLEL AERO' project.

1. Introduction

One of the common problems for application scientists is the porting of codes from serial to parallel computers. Since most of the existing software has been developed on serial computers, exploiting application parallelism without totally rewriting or redesigning existing codes and algorithms is crucial if parallel computers are to be accepted and used by industry. We believe that the availability of portable and efficient parallel numerical libraries, which can be used as building blocks, is extremely important for both simplifying application software development and improving reliability.

Our goal here is to show that combining an incremental methodology with the use of building blocks can greatly simplify the design of mathemat-

P. Wesseling (ed.), High Performance Computing in Fluid Dynamics, 97–144.
© *1996 Kluwer Academic Publishers.*

ical software and the porting of industrial codes while providing satisfying
execution rates. We use as an example the development of codes for parallel
computers for the solution of both full and sparse linear equations and the
porting of some industrial codes from the CERFACS partners Aérospatiale
and CNES, and from the EUROPORT-1 'PARALLEL AERO' project.

As users of high-performance computers, we are involved with different
kinds of tasks of increasing complexity :

- Porting existing codes to parallel computers : only local changes are
  made in the code (use of compiler directives, modifications at loop
  level, ...), there is no change in the global solution technique, except
  possibly the substitution of some computational kernels (linear system
  solution for example). This is the fastest way to obtain some benefits
  of parallelization but it obviously does not lead to the most efficient
  parallelization.
- Developing a parallel code : in this case one has to use a solution
  technique suitable for parallelization that may be very different from
  that used on a serial computer (use of domain decomposition for PDE
  problems for example).
- Developing portable and efficient parallel numerical software : obviously portability and efficiency issues are fundamental and complicate
  the design of the code. For example, the data distribution on a distributed memory machine should not prevent the use of convenient
  data allocation schemes for the user.

Some of the main factors to be considered before porting industrial
applications to parallel computers are the following :

- the suitability of the code for implementation on a parallel computer
  (the solution technique may be inherently serial),
- the scalability of the code (this dictates its suitability for implementation on massively parallel computers),
- the suitability of the application for functional or data parallelism (important issue for implementation on SIMD computers),
- the amount of effort to be spent in parallelizing the code.

Moreover, there is a large variation in programming environments depending on whether the target machine is a shared memory multiprocessor
(where automatic parallelization and loop-level parallelism are available),
a distributed memory multiprocessor where message passing or less often
virtual shared memory are available, or an SIMD computer.

The increasing power of workstations and the evolution in scientific
computing from central computing to distributed computing make very
attractive the use of networks of computers to solve large-scale problems.
This is a very cost-effective opportunity since it allows to use the idle cycles

of computers (often workstations) that potentially provide the processing power of a supercomputer –and also memory and amount of disk storage– at a cost orders of magnitude below (Hariri and Varma (1993)). Additionally, the workstations often use the same off-the-shelf RISC processors, and network of workstations the same programming model – message-passing using for example P4, PVM, or MPI – than MPPs.

The performance of distributed applications depends to a large extend on the communication requirements and on the bandwidth and latency of communications among the network computers. This clearly determines the applicability of distributed computing. The speed of networks such as Ethernet or even or FDDI restricts the use of heterogeneous computing to applications that do not require a large amount of communications compared to computations.

These lecture notes are based on material that results from collaborations between researchers from the Parallel Algorithms Group at CERFACS and the Parallel Algorithms and Optimization Group at ENSEEIHT-IRIT (Daydé and Duff (1995)). Part of this work was supported within the ESPRIT III EUROPORT-1 'PARALLEL AERO' project (Van Kemenade, Daydé, and Vos (1995)),) and also by Aérospatiale, Division Avions and Centre National d'Etudes Spatiales under CERFACS contracts 11C05770 and 873/CNES/90/0841/00. The lectures notes on distributed and heterogeneous computing using PVM are based on the joint ENSEEIHT-IRIT and CERFACS course.

We consider the porting of industrial codes on high-performance computers in Section 2. We then give in Section 3 a quick description of the PVM computing environment for distributed and heterogeneous computing. In Section 4, we give some examples of parallelization of industrial codes, and we conclude in Section 5.

2. Porting industrial codes on high-performance computers

2.1. INTRODUCTION

Todays high-performance computers can be divided into two classes :

- General scientific computers : these are the conventional high performance computers capable of handling efficiently a wide range of applications. Automatic parallelization is generally available, and they support multiprogramming and time-sharing (RISC workstations, shared memory computers with a moderate number of processors including minisupercomputers and vector supercomputers).

- Massively parallel computers : they are characterized by a more limited applicability compared to the current vector supercomputers for example. This typically arises from novel features in the architecture or from limitations in the software (application software, programming tools, ...). They are very competitive in peak performance with the conventional supercomputers and may achieve an interesting price to performance ratio on suitable applications. This class of computers includes both physically distributed memory architectures (both virtual shared memory computers and explicit message passing architectures) and SIMD computers.

Some of the main architectural factors influencing the performance of a computer are the following (Schreiber and Simon (1992)) :

- the memory bandwidth and the latency for accessing data in the memory,
- the relative cost of computing and communication (for remote access on a virtual shared memory computer or for exchanging a message on a multicomputer),
- the synchronization costs,
- the vector start-up time for a vector processor based architecture.

Note that these are only some of the main factors. For example, the I/O requirements of an application are often ignored, while they may be one of the main bottlenecks in performance (see the experiments on the Teledetection application). Any mismatch in the design of an architecture will penalize the application performance irrespective of the efforts of the algorithm designer.

Some other factors that may affect the performance are application dependent :

- The type of parallelism that can be exploited within the application. It can be either functional or data parallelism.
- Task granularity.
- Scalability which imposes limitations on the number of processors that can be efficiently exploited.
- Load balancing.
- Communication needs which may have a great impact on the performance. Locality of references to data is crucial for efficient implementation on distributed memory architectures.
- Regularity of treatment is important for the vector performance and for efficient implementation on SIMD computers.

Clearly, the solution technique has a great impact on these factors (especially on the global and local communication requirements). Some of the basic techniques that are often used when designing software for parallel computers are:

- partitioning and/or blocking (for example, domain decomposition, blocked LU factorization)

- reuse of data (for example, use of loop unrolling or explicit copying)
- exploitation of multiple levels of parallelism (see forthcoming example of sparse solution techniques)
- use of building blocks (for example, use of BLAS)

2.2. HIGH-PERFORMANCE CPUS

Vector processors are commonly used in supercomputers. Recently very fast RISC processors which can also process vectors efficiently have come on to the market. They are usually more efficient than vector processors on scalar applications. The main reasons for their success in the marketplace is their very good cost to performance ratio. They are used as a CPU both in workstations and in most of the current MPPs (DEC Alpha on CRAY T3D, SPARC on CM5 and PCI CS2, HP PA on CONVEX EXEMPLAR, and RS/6000 on IBM SP1 and SP2). We report in Table 1 the uniprocessor performance of some current RISC processors on the double precision 100-by-100 and 1000-by-1000 LINPACK benchmarks (Dongarra (1992)). We also record their peak performance.

Computer	LINPACK 100*100	LINPACK 1000*1000	Peak performance
DEC 8400 5/300	140.0	411.0	600
IBM POWER2-990	140.0	254.0	286
HP 9000/755	41.0	107.0	200
SGI POWER challenge	104.0	261.0	300

TABLE 1. Performance in MFlops of RISC processors on the double precision LINPACK benchmarks

2.3. USE OF BUILDING BLOCKS

As we have previously discussed, it is very important to use standard building blocks when designing or porting codes. They are extremely useful for simplifying the design of codes while guaranteeing portability and efficiency. Two of the most important such building blocks are Fast Fourier Transforms (FFTs) and Basic Linear Algebra Subprograms (BLAS). We consider the effect of using BLAS in the solution of linear equations, first when the coefficient matrix is full, then when it is sparse. We also make heavy use of building blocks such as the BLAS in the case of porting industrial codes.

Different levels of BLAS are available. The level terminology arises from the fact that if the vectors and matrices involved are of order N, the Level 1

BLAS provides vector computations of order $O(N)$ (Lawson, Hanson, Kincaid, and Krogh (1979a), Lawson, Hanson, Kincaid, and Krogh (1979b)), the Level 2 BLAS provides matrix-vector computations of order $O(N^2)$ (Dongarra, Du Croz, Hammarling, and Hanson (1988)), and the Level 3 BLAS provides matrix-matrix computations with $O(N^3)$ operations (Dongarra, Du Croz, Duff, and Hammarling (1990a, 1990b)).

The architecture of high-performance computers usually involves a memory hierarchy (main memory, local memory, vector and scalar registers, ...). All the arithmetic computations are performed at the top level of this hierarchy. Therefore the key to efficiency is to keep active data as close as possible to the top of hierarchy. The use of higher Level BLAS provides this capability. The increased granularity of higher Level BLAS also allows more efficient parallelization depending on the synchronization overheads (Daydé, Duff, and Petitet (1992)). We report in Table 2 the performance of some manufacturer-supplied routines from the BLAS using 64-bit arithmetic (DAXPY, DGEMV and DGEMM on ALLIANT and IBM, and SAXPY, SGEMV, SGEMM on the CRAY-2). A tuned manufacturer-supplied version of the BLAS is today available on most high-performance computers and, if not, a standard Fortran implementation is available on the netlib electronic server (Dongarra and Grosse (1987)). Also a tuned subset of the BLAS is publically available via anonymous ftp at ftp.enseeiht.fr in pub/numerique/BLAS/RISC (Daydé and Duff (1996)).

Computer	Level 1 BLAS _AXPY	Level 2 BLAS _GEMV	Level 3 BLAS _GEMM
ALLIANT FX/80 (1 proc.)	3.3	4.0	13.0
ALLIANT FX/80 (8 proc.)	17.5	31.9	89.4
CRAY-2 (1 proc.)	121.0	316.5	437.5
IBM 3090E/VF (1 proc.)	26.0	60.0	80.0

TABLE 2. Performance in Mflops of BLAS routines on some computers

The BLAS can be successfully used in designing codes for the solution of the linear system

$$Ax = b. \tag{1}$$

We first discuss the case when A is full, then when A is sparse.

2.3.1. Solution of full linear systems

There are several ways to design full linear solvers using BLAS. One algorithm is to factorize the matrix by blocks of columns, using Level 2 BLAS within the block which has previously been updated by all the previously factorized matrix using the Level 3 BLAS routines TRSM and GEMM.

We do not want to go into the details of the implementation here but refer the reader to Daydé and Duff (Daydé and Duff (1991)) for further information. We show in Table 3 a summary of results obtained by Daydé and Duff (Daydé and Duff (1991)) where, for each machine, the same code, all in Fortran, has been used. This is clearly at least competitive with the manufacturer's Library, normally written in assembler. We also compare in that table two types of parallelization : in the Parallel BLAS version, the parallelism is only exploited within the Level 3 BLAS, while in the Parallel LU version, additional freedom is obtained by parallelization over the BLAS. The use of parallel BLAS kernels is more and more common for exploiting parallelism while maintaining portability. Of course, this approach captures only part of the potential parallelism. The Parallel LU version is more efficient at the price of a decreased portability.

Computer	Procs	Block LU factorization			Manufacturer's Lib.	
		Serial	Parallel BLAS	Parallel LU	Routine	Perf.
ALLIANT FX/80	8	12	65	62	PDGEFA	39
CRAY-2	4	379	1027	1072	SGEFA	353
IBM 3090E-400	3	64	132	183	DGEF	72
IBM 3090J-600	6	89	294	418	DGEF	97

TABLE 3.   Performance in MFlops of the LU factorization on shared memory multiprocessors using 1000-by-1000 matrices

The LAPACK library (Anderson, Bai, Bischof et al. (1992)) uses as much as possible block algorithms to take advantage of the higher level of BLAS. It can solve systems of linear equations, linear least squares problems, eigenvalue problems, and singular value problems. It is designed to give high efficiency on vector processors, RISC-based computers, and shared memory multiprocessors.

The ScaLAPACK library (Choi, Demmel, Dhillon et al. (1995)) is an extension of LAPACK for distributed memory computers. It illustrates how important is this notion of building blocks for reusing as much as possible existing software and developing portable and efficient codes. ScaLAPACK is based on the BLAS and LAPACK libraries. Two additional libraries have been introduced : the BLACS (Basic Linear Algebra Communication Subprograms, see Dongarra and Whaley (1995)) that are used as a communication layer (on top of message passing libraries such as PVM, NX, MPL,

CMMD,..), and the PBLAS (Parallel Basic Linear Algebra Subprograms, see Choi, Dongarra, Ostrouchov et al. (1995)).

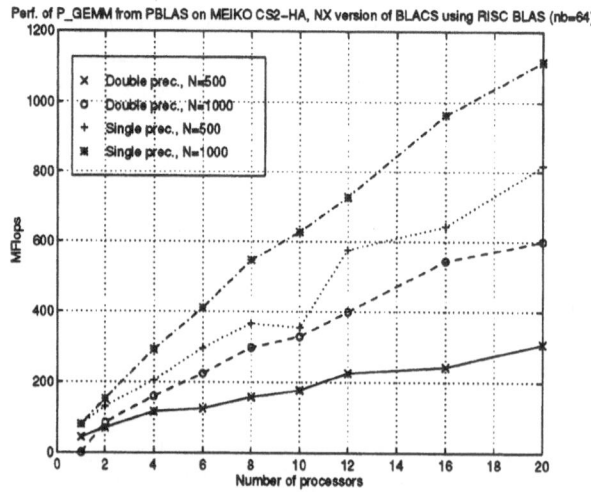

Figure 1.   Performance of parallel matrix-matrix product from PBLAS on MEIKO CS2-HA

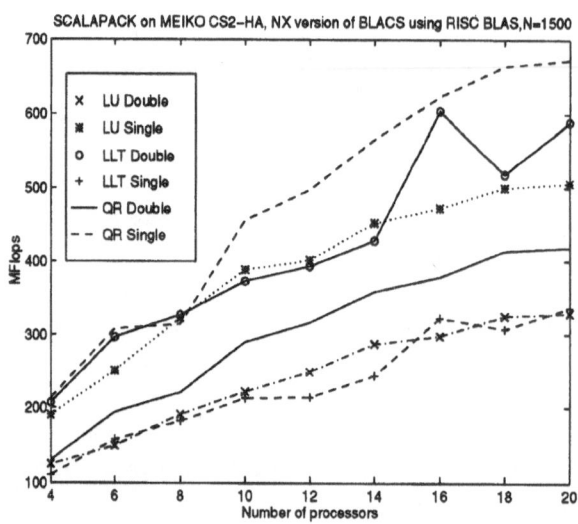

Figure 2.   Performance of dense factorizations from ScaLAPACK on MEIKO CS2-HA

We report in Figure 1 the performance of the parallel matrix-matrix product from PBLAS (we consider single and double precision kernels i.e. respectively PSGEMM and PDGEMM) on the MEIKO CS2-HA installed at CERFACS, using square matrices of order 500 and 1000. The processor is a 100 Mhz HyperSPARC of peak performance equal to 100 MFlops. We use the version of BLACS on top of the NX message passing library. The tuned serial BLAS used is the RISC BLAS (Daydé and Duff (1996)) which is available on anonymous ftp. We also report in Figure 2 the performance of the LU, Cholesky and QR factorizations from ScaLAPACK version 1.0 using both single and double precision on relatively small matrices of order 1500.

2.3.2. *Solution of sparse linear systems*

We consider the sparse LU factorization of square matrices on shared memory multiprocessors. We use a multifrontal approach to design our factorization algorithms. Further background on this approach can be obtained from the original papers by Duff and Reid (1983, 1984).

In a multifrontal method, the sparse factorization proceeds by a sequence of factorizations on small dense matrices, called frontal matrices. The ordering for the sequence of computations and the frontal matrices are determined by a computational tree, called elimination tree, where each node represents a full matrix factorization and each edge the transfer of data from child to parent node. This elimination tree is determined from the sparsity pattern of the matrix and from a reordering that aims at minimizing the fill-in during the numerical factorization (such as the minimum degree we use here). During the numerical factorization, eliminations at any node can proceed as soon as those at the child nodes have completed. This will be referred to as tree parallelism.

Note that the factorization at each node is done using full linear algebra and direct addressing so that we can use the BLAS. All the indirect addressing is confined to the assembly process.

We have developed a parallel multifrontal code for the solution of symmetrically structured unsymmetric equations. The results in this section are from runs with this code, called **MUPS**, and experimental versions of this code (Amestoy and Duff (1989), Amestoy (1991), Amestoy and Duff (1993), Amestoy, Daydé, Duff, and Morère (1995)). Recently, a library version of this research code, called **MA41**, has been designed for the Harwell Subroutine Library Release 12.

During the LU factorization, if we only exploit the tree parallelism, the speed-up is very disappointing. The actual speed-up depends on the problem but is typically only 2 to 3 irrespective of the number of processors. This poor performance is caused by the fact that the tree parallelism decreases while going towards the root of the tree. Moreover, Amestoy and Duff Amestoy and Duff (1993) have observed that typically 75% of the work is performed in the top three levels of the elimination tree. It is thus necessary to obtain further parallelism within the large nodes near the root of the tree (so-called node parallelism) by using parallel versions of the

BLAS in the factorizations within the nodes. When combining both tree and node parallelism the situation becomes much more encouraging and we show typical speed-ups for a range of computers in Table 4. A medium size sparse matrix, BCSSTK15 from the Harwell-Boeing set (Duff, Grimes, and Lewis (1992)), is used to illustrate our discussion. This is a structural analysis matrix of order 3948 with 117816 nonzeros. A minimum degree ordering is used in the analysis and the number of floating-point operations for the factorization is 443 million.

Computer	nprocs	(1) Mflops	(speed-up)	(2) Mflops	(speed-up)
Alliant FX/80	8	15	(1.9)	34	(4.3)
IBM 3090E/3VF	3	83	(1.9)	105	(2.4)
IBM 3090J/6VF	6	126	(2.1)	227	(3.8)
CRAY-2	4	316	(1.8)	404	(2.3)
CRAY Y-MP	6	529	(2.3)	1119	(4.8)

Table 4: Performance summary of the multifrontal LU factorization on matrix BCSSTK15. In column (1) we exploit only parallelism from the tree; in column (2) we combine the two levels of parallelism.

2.4. METHODOLOGY FOR PORTING CODES TO PARALLEL COMPUTERS AND PORTABILITY

In most of our studies on porting codes, we have used an incremental approach (Charles, Daydé, Petitet et al. (1993), Daydé, Duff, L'Excellent, and Giraud (1993)). First, the code is parallelized on shared memory multiprocessors (where compilers have automatic vectorization and parallelization capabilities). Then, on the basis of the scalability and data locality properties of the code, we decide whether to port the application to distributed memory computers (virtual shared or message passing environment) or even to networks of computers. This methodology for porting codes can be described as follows :

- **Step 0** : Start with a portable code (standard Fortran or C)
- **Step 1** : Use automatic vectorization and parallelization. Profile the code in order to identify the most-time consuming parts of the code.
- **Step 2** : Clean up the code. Use loop level tuning and tuned computational kernels.
- **Step 3** : Manual parallelization of the code. This may require changing parts or all of the solution technique.
- **Step 4** : Decide whether it is sensible to port the code to massively parallel computers or to networks of workstations.

The gains from cleaning up a code and using appropriate compilation directives may be significant, especially on codes written before the availability of vector computers. Cleaning up a code may involve the following operations :

1. Suppression of some subroutine calls
2. Tuning the vectorization and the parallelization at loop-level :

   – Switch some loops (parallel outer, vectorial inner)

   – Precompute the loop invariants

   – Check and possibly change the dimensions of the arrays

3. Check and possibly modify I/O
4. Identification of building blocks (BLAS, FFT, ...)

2.4.1. *Portability problems*

Since standards both for Fortran and C exist, it should be easy to write portable code. However, in practice, problems often arise when porting codes. They may come from poor or non-standard implementations of Fortran or C by the vendor, but often they come from the design of the code (for example use of binary files).

The most frequent portability problems are the following :

– Use of non-standard Fortran or C
– I/O portability problems : use of binary files, non-standard I/O procedures (random I/O,...)
– Non IEEE arithmetic (for example on CRAY and CDC computers)
– Implicit typing of variables

The use of different arithmetics by different vendors may cause noticeable differences in the number of iterations required for convergence of iterative methods. We consider the solution of a linear system used in a 3D transonic aerodynamic code (Daydé, Duff, L'Excellent, and Giraud (1993)). The number of iterations required for convergence of a preconditioned conjugate gradient code and the residuals on three computers are given by :

1. CRAY XMP : 15 iterations, residual $= 0.21718 \times 10^{-5}$
2. CRAY 2 : 17 iterations, residual $= 0.61356 \times 10^{-6}$
3. CONVEX C220 : 15 iterations, residual $= 0.88515 \times 10^{-5}$

It is interesting to note the influence of the arithmetic on the convergence rate and on the precision achieved.

2.4.2. *Fortran developments*

Some new derivatives of Fortran or extensions to Fortran are available (Fortran 90, Fortran D, High Performance Fortran, and the Parallel Computing

Forum extensions) that incorporate features to aid in the exploitation of parallel computers :

- Arrays extensions for data parallelism
- Parallel constructs (such as parallel DO-loop)
- Data distribution directives for distributed memory computers (including SIMD)

Fortran 90 incorporates many new features (Fortran 90 (1991)). The main one of interest here is array extensions. For example, it is possible to write : $A = B + C$ where A, B, and C are arrays. There also exists new intrinsic functions that are applicable to arrays. There is obviously scope for exploiting data parallelism within these operations. This means that the underlying data parallelism of Fortran 90 could possibly be efficiently supported on architectures such as shared memory architectures and some SIMD computers (MasPar and Thinking Machine computers have provided several Fortran 90 constructs for some time).

Since data distribution is crucial on multiprocessors where the latency for accessing remote data is high, it can be seen that the main drawback of Fortran 90 is that it does not offer ways of specifying an efficient data distribution.

This is the reason why High Performance Fortran, which is a superset of Fortran 90, was initially proposed (High Performance Fortran Forum (1993)). It incorporates data distribution directives similar to those available within Fortran D. HPF Fortran also provides the FORALL statement that can be used to specify identical sequences of operations on elements of arrays permitting simultaneous execution.

The data programming style, already used on SIMD computers, can obviously be supported on MIMD computers. It often scales up to large numbers of processors and can be used on many applications.

Parallelizing applications using message passing is surprisingly more portable since message passing can be emulated on shared and virtual shared memory computers, and since it is the programming paradigm used on multicomputers and networks of computers. Some message passing libraries are available on a wide range of computers such as PVM (Geist, Beguelin, Dongarra et al. (1993)), and P4 (Butler and Lusk (1992)). An international collaboration has defined a standard for message passing : MPI (Dongarra, Hempel, Hey, and Walker (1995)).

3. Distributed and heterogeneous computing using PVM

We first give, in Section 3.1, a general overview of the PVM (Parallel Virtual Machine) computing environment. We describe, in Section 3.2,, the main features of the PVM user Library. In Section 3.3, we then illustrate the use of PVM on a straightforward implementation of a simple example. The XPVM tracing tool will be used to show that our straightforward implementation cannot take heterogeneity into account. A modified implementation of our test example will be presented in Section 3.4 to illustrate the impact of dynamic scheduling on the behaviour of a parallel application.

For a complete description of the PVM environment, we advise the reader to consult Beguelin et al. (1991, 1993) and Geist et al. (1993). Many of the descriptions provided in this document are based on the previously mentioned documents.

3.1. OVERVIEW OF THE PVM COMPUTING ENVIRONMENT

PVM is a public domain software system available on *netlib* and developed by the Oak Ridge National Laboratory, the University of Tennessee, the University of Carnegie Mellon, the Pittsburgh Supercomputing Center and the Emory University of Atlanta. This programming environment allows to use a network of heterogeneous UNIX computers (either serial or parallel) as a unique computing resource referred to as a *virtual machine*. A variety of networks (Ethernet, FDDI, ...) may interconnect the nodes of the virtual machine (see for example Figure 3).

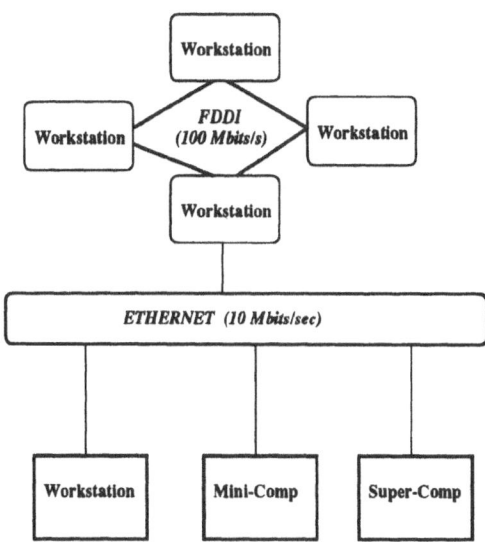

Figure 3. Example of virtual machine

A process daemon, activated on each node of the virtual, coordinates work distributed on the virtual machine. A file containing the list of computers allows to automatically activate the UNIX daemons and therefore

build the parallel virtual machine. The parallel application is then viewed
as a set of parallel processes being executed on the processors of the virtual
machine that communicate and synchronize using message passing programming model. The processes can be organized into groups (a process
can belong to several groups and groups can change at any time during
computation).

From the user point of view, the PVM package is composed of two parts :

- **a daemon process**

  A daemon, called *pvmd3*, resides on each computer of the parallel virtual machine. The daemon can be started interactively or automatically. When a user wants to run an application in the PVM environment, he first launches a procedure (*pvmd3*) which can automatically
  start a daemon on each node of a virtual machine described in a *host
  file*. The parallel application can then be started from any node of the
  virtual machine computer. The command *pvm* starts the PVM console
  used to interactively control and modify the virtual machine both in
  terms of host nodes and processes. The control command *pvm* may be
  started and stopped multiple times on any of the hosts.

- **a set of library procedures**

  The PVM3 library contains communication and synchronization procedures enabling the user to exploit the virtual machine using programs
  written either in C or in FORTRAN. The library routines provide several facilities for handling 'processes', which are executing instances of
  a program. In particular, PVM provides facilities to create and ter-
  minate processes, to communicate between processes, to synchronize
  processes, to modify the parallel virtual machine, and to manipulate
  process groups.

3.2. THE PVM3 USER LIBRARY

For the sake of simplicity, we focus, in this section, on the description of
the main procedures of the FORTRAN PVM user library. For a complete
version of PVM3 user library we advise the reader to consult the "PVM 3
user's guide and reference manual" Geist, Beguelin, Dongarra et al. (1993)
available on netlib by email (netlib@ornl.gov). More advanced and recent
features of PVM3 are overviewed in Beguelin, Dongarra, Geist et al. (1995).

We first describe the primitives that perform control and creation of processes. We then present the interprocess communication and process group
manipulation primitives. In Section 3.3, we illustrate the most commonly
used procedures of PVM3 on a very simple example.

Remark:
To use the predefined options and the error message coding, the file fpvm3.h
must be included in the FORTRAN code
(include '/usr/local/pvm3/include/fpvm3.h').

3.2.1. *Control and Activation of processes*
- **Procedure for enrolling a process into PVM**
    call **pvmfmytid**(tid)
    At its first call, the *pvmfmytid()* procedure creates a PVM process.
    pvmfmytid() returns the process identifier *tid* and may be called several
    times. If the host node does not belong to the parallel virtual machine
    then an error message is returned.

- **Leave PVM**
    call **pvmfexit**(info)
    pvmfexit indicates to the local daemon (*pvmd3*) that the process leaves
    the PVM environment. The process is not killed but it cannot anymore
    communicate (via PVM communication procedures) with the other
    PVM processes.

- **Kill another PVM process**
    call **pvmfkill**(tid, info)
    pvmfkill kills the PVM process identified by *tid*.

- **Starting other processes on the virtual machine**
    call **pvmfspawn**(task, flag, where, ntask, tids, numt)
    pvmfspawn starts *ntask* copies of the executable file *task*. The choice
    of the parameter *flag* allows to control the type of computer on which
    will be activated the processes.

    Predefined values of *flag* :
PvmDefault	PVM chooses the computers on which the processes will be activated.
PvmArch	the parameter *where* then defines a target architecture.
PvmHost	the parameter *where* then specifies a target computer.
PvmDebug	processes are activated in debugging mode.

The parameter *numt* holds, on exit, the number of PVM processes
actually activated. The corresponding task identifiers are located the
first *numt* positions of the array *tids(ntask)*. The error codes (negative
returned values) corresponding to non-activated processes are located
in the last *ntask-numt* positions of the array *tids()*.

- **Getting the tid of the parent**

call **pvmfparent**(tid)

On exit *tid* holds the task identifier of the parent process of the calling process. If the calling process has not been created by a call to *pvmfspawn* then *tid* is set to the negative value *PvmNoParent*.

3.2.2. *Interprocess communication*

The communication between PVM processes is based on the message-passing programming model. PVM provides asynchronous send, blocking receive, and nonblocking receive facilities.

Sending a message is effectively done in three steps (see Figure 4):

1. initialization of a send buffer and choice of an encoding format to send data; (*pvmfinitsend*)

2. packing of data to be sent into the send buffer (*pvmfpack*);

3. actual send/broadcast of the message stored in the send buffer to destination(s) process(es) (*pvmfsend, pvmfcast*).

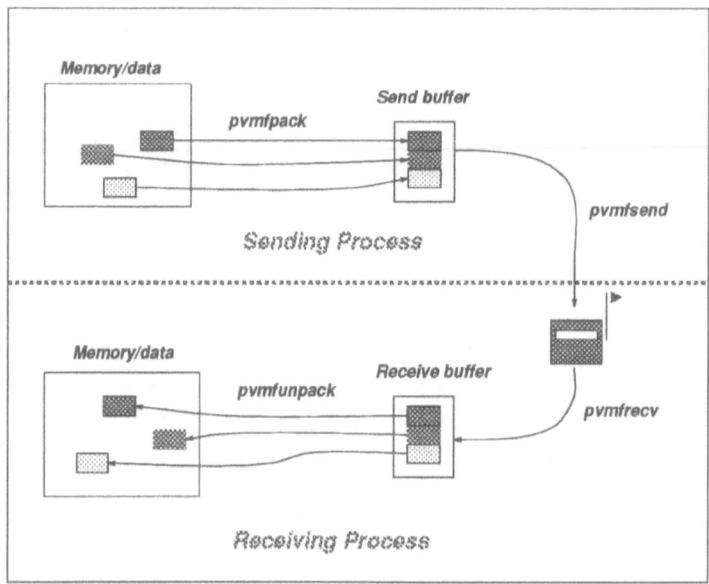

Figure 4. Illustration of send/receive main steps

The main advantage of this strategy is that the user can compose his message out of various pieces of data and therefore decrease the number of messages effectively sent. Furthermore with the broadcast option, only one send buffer has to be filled. Moreover, very often, one large array of only a given data type needs to be sent. To improve the performance of this special

case of point-to-point communication a PVM3 primitive (*pvmfpsend*) has
been designed to pack and send data in one call.

The reception of a message is symmetric to the three step sending procedure (see Figure 4). After reception of a message into the active buffer,
the data are unpacked into the destination arrays.

Various options to receive data are provided:

pvmfrecv	:	blocking receive
pvmftrecv	:	timeout receive
pvmfnrecv	:	nonblocking receive
pvmprecv	:	combines blocking receive and unpacking.

- **Management of buffers:**

Clear/initialize send buffer

call **pvmfinitsend**(encoding, bufid)
pvmfinitsend clears the send buffer and prepare it for packing a new
message. The encoding scheme used during data packing is defined by
the parameter *encoding*.

Predefined values of *encoding* in FORTRAN:

PvmDefault	The XDR encoding is used. It is suitable to data exchange on an heterogeneous network of computers.
PvmRaw	No encoding, the message are sent using the native format of the host node.
PvmInPlace	Data are not copied into the buffer which only contains the size and pointers to the data.

It is also possible to manipulate simultaneously several buffers even if at a
given time there is only one active buffer for sending/receiving data. Procedures to create/release buffers (*pvmfmkbuf, pvmffreebuf*) to get/set the
active send/receive buffer (*pvmfgetrbuf, pvmfgetsbuf, pvmfsetsbuf, pvmfsetr-
buf*) are designed for this purpose (for further details see Geist, Beguelin,
Dongarra et al. (1993)).

- **Packing/unpacking data:**

call **pvmfpack**(what, xp, nitem, stride, info)
pvmfpack packs an array of data of a given type into the active send
buffer. A message containing data of different types may be built using successive calls to *pvmfpack*. *nitem* elements chosen each *stride*
elements of the linear array *xp* of type *what* are packed into the buffer.

Predefined values of *what* :

STRING, BYTE1, INTEGER2, INTEGER4

REAL4, REAL8, COMPLEX8, COMPLEX16

call **pvmfunpack**(what, xp, nitem, stride, info)
Similarly, *pvmfunpack* is used to unpack informations held into the active receive buffer. The unpacked data are then stored into the array *xp*.

- **Sending/Receiving messages:**

call **pvmfsend**(tid, msgtag, info)
pvmfsend sets the message label to *msgtag* then sends it to the pvm process of number *tid*.

call **pvmfmcast**(ntask, tids, msgtag, info)
pvmfmcast broadcast the message to *ntask* processes specified into the integer array *tids*.

call **pvmfnrecv**(tid, msgtag, bufid)
pvmfnrecv performs a non-blocking receive. If the message of label *msgtag* issued by process *tid* is not arrived then *bufid = 0*, otherwise the message is stored into a new buffer *bufid* automatically created. If *tid =-1* then the first message with label *msgtag* from any process will be received. If *msgtag =-1* the label is ignored.

call **pvmfrecv**(tid, msgtag, bufid)
pvmfrecv blocks the process until a message with label *msgtag* has arrived from *tid*. The other functionalities are similar to those of *pvmfnrecv*.

Check for arrived messages

call **pvmfprobe**(tid, msgtag, bufid)
If the message is not arrived then *bufid =0*, otherwise a buffer number is returned but the message is not received.

call **pvmfbufinfo**(bufid, bytes, msgtag, tid, info)
pvmfbufinfo returns the characteristics of the message stored in *bufid*: label *msgtag*, sending process *tid*, length in bytes *bytes*. *pvmfbufinfo* is particularly useful in conjunction with *pvmfprobe* or when the label –or the source– of the message to be received have not been specified.

3.2.3. *Management of process group*

The procedures for managing process groups form a layer on top of the PVM layer. They are provided into a separated library **libgpvm3.a**. A group server (**pvmgs**) is automatically activated at the first called to a procedure of the libgpvm3.a library.

Main characteristics of PVM groups:
- Any PVM process can join a group;
- A process can belong to several groups;
- A message can be broadcasted to a PVM group from any PVM process
- Synchronization within a group can be performed using barriers.

3.2.4. *Performance of PVM*
Manufacturers (IBM, CRAY ...) often provide a tuned implementation of the PVM communication library which make use of the native communication calls, the shared memory or the virtual shared memory. In this way, the user still have a portable PVM based parallel code which automatically exploits the most advanced communication tools available on a target computer. The performance across networks of computers has also been improved by the use of Unix domain sockets between the tasks and the local daemon (improvement by a factor of 1.5 to 2). Also the use of task-to-task direct communications with the *PvmRouteDirect* option increases the communication performance. Note that it is not a scalable solution since the establishment of a TCP link requires a file descriptor.

3.3. ILLUSTRATIVE EXAMPLE: A DOT VERSION OF THE MATRIX VECTOR PRODUCT

We illustrate the use of PVM on the parallelization of a matrix-vector product on a network of workstations. After a short presentation of the parallel algorithm, we give the PVM code of a static parallel implementation. We then describe how to compile the code, define the parallel virtual machine, and finally execute the parallel code on the virtual machine. A dynamic implementation of the matrix vector product will be presented Section 3.4.

3.3.1. *Description of the parallel algorithm*
We describe in this section a straightforward static parallelization of the matrix-vector product,

$$y = A \times x$$

where A is a $(m \times n)$ matrix, x an n-vector, and y an m-vector. We use a master-slave paradigm. Each process is in charge of computing one block of the vector y. Additionally the master process broadcasts the data to the slaves and collects the final results.

The sequential FORTRAN code is:

```
do i = 1, m
  y(i) = 0.0D0
  do j = 1, n
    y(i) = y(i) + A(i,j) * x(j)
  enddo
enddo
```

In the parallel implementation, the matrix A is partitioned into block of rows that will be distributed on the slave processes.

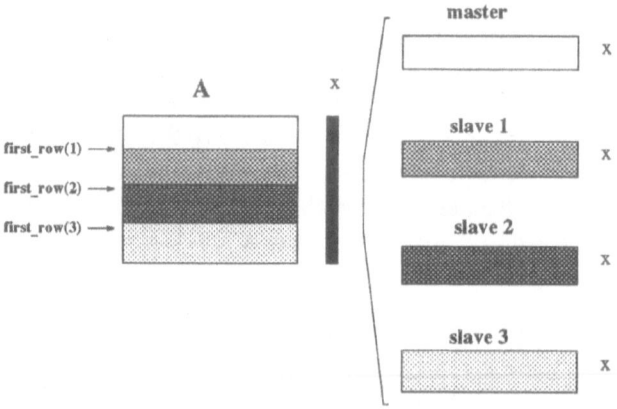

Figure 5. Static parallelization of the matrix vector product.

Description of

• master process	• slave process
(the master holds A and x)	
enroll into PVM and create slaves	enroll into PVM
send data to the slaves	wait for data from the master
compute the first block of y	compute my block of y
receive results from slaves	send back results to master
leave PVM application	leave PVM application

The computation of one block of vector y is effectively performed using a generalized version of the matrix-vector product $y = \alpha A x + \beta y$ available in the _GEMV subroutine from the Level 2 BLAS (see Dongarra et al. (1988)).

3.3.2. *Codes for master and slave processes*

Master's code:

```
      PROGRAM  dotmatvec
      integer slave_max, lda
      parameter (slave_max=32, lda=1000)
* PVM variables
      integer my_id, info, inst(slave_max), numt, bufid
     *        nb_of_slaves, no_slave, nb_of_processes, type,
     *        retcode, first_row(slave_max)
*
* Message types used:
* type = 0 to broadcast initial informations
*      = 1 to distribute data to the slaves
*      = 2 to receive results from the slaves

* Data declaration
      double precision a(lda,lda),x(lda),y(lda),one,zero
      integer          incx,n,m,i,j
      data             zero/0.0/, one/1.0/
      include '/usr/local/pvm3/include/fpvm3.h'

*  Enroll this program into PVM
      call pvmfmytid (my_id)

* read input data (nb_of_slaves, m, n)
      read(*,*) nb_of_slaves, m, n

*  initiate nb_of_slaves instances of slave program
      call pvmfspawn ('slave',PVMDEFAULT,'*',nb_of_slaves,inst,numt)
      if (numt .ne. nb_of_slaves) stop
      nb_of_processes = nb_of_slaves +1

* Initialize data for computation and compute first_row(slave_no)
      do j=1,n
        do i = 1,m
          a(i,j) = DBLE(i+j)/DBLE(m+n) + one
        enddo
        x(j) = one + DBLE(j)/DBLE(n)
      enddo
      j = (m / nb_of_processes)
      do i=1,nb_of_slaves
        first_row(i) = i*j +1
      enddo
      first_row(nb_of_processes) = m+1
*     work balancing
      j = mod(m,nb_of_processes)
      do i=1, j-1
        first_row(nb_of_processes-i) =
```

```
      &           first_row(nb_of_processes-i) +j -i
          enddo

*   broadcast the number of columns and x to each slave process
          type = 0
          call pvmfinitsend (PvmDefault, bufid)
          call pvmfpack (INTEGER4, n, 1, 1, info)
          call pvmfpack (REAL8,    x, n, 1, info)
          call pvmfmcast (nb_of_slaves, inst, type, info)

*   send its sub-matrix data to each slave process
          type = 1
          do 60, no_slave = 1, nb_of_slaves
*          number of components computed by slave  no slave
             j = first_row(no_slave+1) - first_row(no_slave)
*          initialization of send buffer
             call pvmfinitsend (PvmDefault, bufid)
*          pack data into send buffer
             call pvmfpack (INTEGER4, j, 1, 1, info)
             do 70, i=1, n
                call pvmfpack (REAL8,a(first_row(no_slave),i),j,1,info)
 70          continue
*          send message stored in send buffer to slave  inst(no slave)
             call pvmfsend (inst(no_slave), type, info)
 60       continue

*      compute its part of the work
*      perform y <-- one*Ax + zero*y
*      where A is an matrix of order  (first row(1)-1) x n.
          incx = 1
          call dgemv('N',first_row(1)-1,n,one,a,lda,x,incx,zero,y,incx)

*      collect results of slave processes and quit PVM
          type = 2
          do 80, no_slave = 1, nb_of_slaves
*             j = number of components computed by the slave no_slave
             j = first_row(no_slave+1) - first_row(no_slave)
             call pvmfrecv (inst(no_slave), type, bufid)
             call pvmfunpack (REAL8,y(first_row(no_slave)),j,1,info)
 80       continue
          call pvmfexit (retcode)
          stop
          end
```

Slave's code:

```
**************************************************************
      PROGRAM slave
*
      include '/usr/local/pvm3/include/fpvm3.h'
      integer from_tid, p_id, bufid, type, recvlen,
     *        my_id, info, lda
      parameter(lda=1000)
      double precision a(lda,lda),x(lda),y(lda),one,zero
      integer incx,n,m,i
      data    zero/0.0/, one/1.0/
```

* **Enroll this program in PVM`3**
```
      call pvmfmytid (my_id)
```

* **Get the tid of the master's task id**
```
      call pvmfparent (p_id)
```

* **receive broadcasted data: number of columns and vector x**
```
      type = 0
      call pvmfrecv (p_id, type, bufid)
      call pvmfunpack (INTEGER4, n, 1, 1, info)
      call pvmfunpack (REAL8   , x, n, 1, info)
*
```
* **receive my block of rows**
```
      type = 1
      call pvmfrecv (p_id, type, bufid)
      call pvmfunpack (INTEGER4, m, 1, 1, info)
      do 10, i=1,n
        call pvmfunpack (REAL8, a(1,i), m, 1, info)
  10    continue
```

* **perform matrix-vector on my block of rows**
```
      incx = 1
      call dgemv('N',m,n,one,a,lda,x,incx,zero,y,incx)
```

* **send back results to master process**
```
      type = 2
      call pvmfinitsend (PVMRAW, bufid)
      call pvmfpack (REAL8, y, m, 1, info)
      call pvmfsend (from_tid, type, info)
```

* **leave PVM environment**
```
      call pvmfexit (info)
      stop
      end
```

3.3.3. Compilation - Link

The following makefile shows how to compile and link the previous codes.
The makefile variables *PvmArch* and *PvmDir* correspond respectively to
the target computer and to the localization of the PVM library. These
variables should be set according to the target computer and to the PVM
library installation. We have assumed that the BLAS library is available
(link done with -lblas).

Example of Makefile

```
F77   = /usr/lang/f77
FOPTS = -O -u
*
#
# Specification of the target computer
PvmArch       =       SUN4
#
# Localization of PVM libraries
PvmDir   =       /usr/local/pvm3/lib

# PVM libraries (C, FORTRAN, Group)
PVMLIB_C =       $(PvmDir)/$(PvmArch)/libpvm3.a
PVMLIB_F =       $(PvmDir)/$(PvmArch)/libfpvm3.a
PVMLIB_G =       $(PvmDir)/$(PvmArch)/libgpvm3.a
LIBS     =       $(PVMLIB_F) $(PVMLIB_C) $(PVMLIB_G)
#
# Localization of the executable files
IDIR     = $(HOME)/pvm3/bin/$(PvmArch)
#
all : dotmatvec slave

dotmatvec : master.o $(BLAS) $(TIMING)
        $(F77) -o dotmatvec master.o $(LIBS) -lblas
        mv dotmatvec $(IDIR)

slave : slave.o $(BLAS)
        $(F77) -o slave slave.o $(LIBS) -lblas
        mv slave $(IDIR)

.f.o :
        $(F77) $(FOPTS) -c $*.f

clean :
        /bin/rm *.o
```

We see in the makefile that the executable files have been moved to the
directory *IDIR* . To start both the *pvmd3* daemons and the slave processes,
the corresponding executables must be accessible. A simple solution is to

automatically include in the shell environment variable *PATH* the correponding paths.

3.3.4. *Configuration of the virtual machine*

A configuration file is used to describe the list of computers involved in the PVM application. This file is used to start the *pvmd3* daemon on each computer listed in the configuration file. It is then possible to control the configuration of the parallel machine using the *pvm* console. Another solution is to use directly the *pvm* console to build the virtual machine. Both solutions are illustrated in the following example where a parallel virtual machine of 4 RISC workstations (HP, IBM and two SUN) is built. The HP workstation, *pie*, is our host computer to illustrate how to build a parallel machine.

Example

```
          *** 1/ Use of a hostfile ***

pie> cat hostfile
# comments
pie
pinson
goeland
aigle
pie> pvmd3 hostfile &
pie> pvm
> conf
4 hosts, 1 data format
                HOST     DTID     ARCH    SPEED
                 pie    40000     HPPA     1000
              pinson    80000     SUN4     1000
             goeland    c0000     RS6K     1000
               aigle   100000     SUN4     1000


          *** 2/ Use of the pvm console ***

pie> pvm
pvm> conf
1 host, 1 data format
                HOST     DTID     ARCH    SPEED
                 pie    40000     HPPA     1000
pvm> add pinson goeland aigle
3 successful
                HOST     DTID
              pinson    80000
             goeland    c0000
               aigle   100000
```

3.3.5. *Performance analysis and graphical interface*

The analysis of the efficiency of the parallel execution of a program is a complex problem by itself. Time measures and speed-up estimations are often not sufficient to understand the behaviour of the parallel application. Automatic tracing of the execution of a parallel execution is an indispensable tool for the programmer. Tracing is very useful both to visualize the parallel behaviour of the application and to debug the parallel code.

Although PVM provides a level of flexibility that allows the programmer to control which events are generated and where messages will be sent, most programmers do not want to deal with such enhanced PVM features. XPVM is a tracing tool exploiting automatically the tracing features of PVM. XPVM is based on previous work done on Xab by Beguelin, Dongarra, Geist, and Sunderam (1993) and Beguelin (1993). Since PVM's tracing facilities generate extra traffic in the network, it is important to realize that this traffic will perturb the running characteristics of the program execution.

We show in Figure 6 the traces obtained using XPVM during the parallel execution of the matrix vector product (see Section 3.3). The target virtual machine is an heterogeneous set of four RISC workstations. The master process is located on computer node *rosanna*.

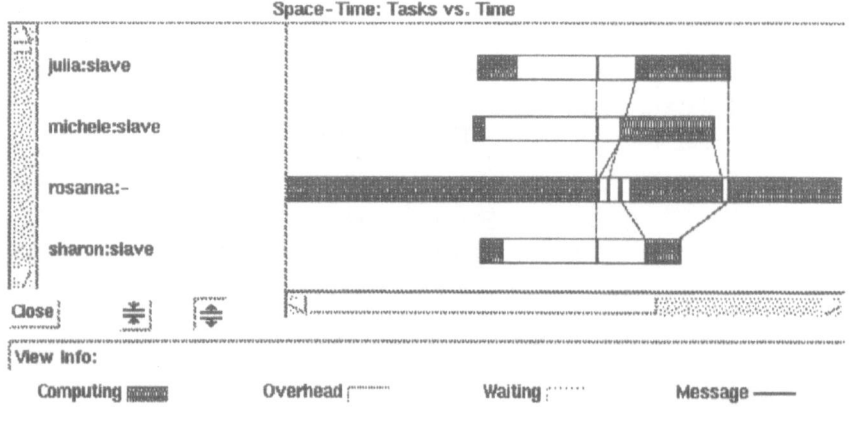

Figure 6.   XPVM trace of the static parallelization of the matrix-vector product

As one could expect from our parallel implementation, we see in Figure 6 that our static partionning of the work does not exploit correctly the potential of the fastest computer (*sharon* on our example) of the parallel virtual machine, which is idle most of the time.

3.4. DYNAMIC SCHEDULING AND HETEROGENEITY

To exploit the heterogeneity of the parallel virtual machine, one could statically build a partitionning of the work taking into account both the heterogeneity of the network and of the processor nodes of the parallel virtual machine. This solution is not always easy to implement, and is not handling the dynamic heterogenity of the parallel virtual machine. In practice, when running a parallel application on a network of workstations (shared by multiple users), the load of the computer/network may vary during execution and between two consecutives executions. The only solution to this problem is to dynamically (at execution time) migrate work on the least loaded or the most powerful node of the virtual machine. This will be referred to as *dynamic scheduling strategy* in the remainder of this paper.

As an illustrative example, we show how dynamic scheduling can be used to parallelize the matrix vector product. The matrix is partitioned into more blocks that the number of slave processes involved (see Figure 7).

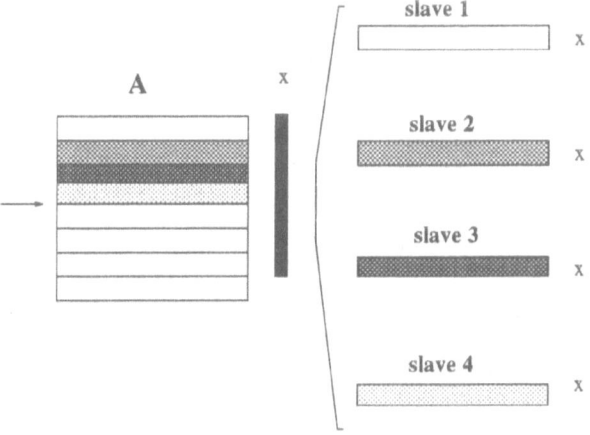

Figure 7. Description of the partitioning used for the dynamic version of the matrix vector product

The master process distributes data and work onto the slave processes. Each time a slave process has finished its currrent block, it sends the corresponding results to the master. It then asks for more work to the master until all the matrix has been processed.

3.4.1. *Codes for master and slave processes*
Master's code:

```
      PROGRAM masterD
* we only show the most important parts of the code.
* (declarations/initialisations are very similar
* to the static implementation and have been suppressed)
*
*
* Enroll this program in Pvm
      call pvmfmytid(my_id)
*
* initialize nb_of_slaves instances of slave program
      call pvmfspawn ('slave',PVMDEFAULT,'*',nb_of_slaves,inst,numt)
*
* nblig : task granularity (number of lines of A processed/task)
* nbtasks: number of blocks to be processed
         nbtasks = m/nblig
* compute the index of the first row performed by each process
      first_row(1) = 1
      do 30, i=1,nbtasks-1
        first_row(i+1) = i*nblig +1
 30    continue
      first_row(nbtasks+1) = m+1
*
* broadcast the number of columns and x to each slave process
      type =0
      call pvmfinitsend( PVMRAW, bufid)
      call pvmfpack(INTEGER4, n, 1, 1, info)
      call pvmfpack(REAL8,    x, n, 1, info)
      call pvmfmcast(nb_of_slaves, inst, type, info)
*
* send its first submatrix to each slave process
      do 60, no_slave = 1, nb_of_slaves
        task_id = no_slave
        j = first_row(task_id+1) - first_row(task_id)
        call pvmfinitsend( PVMRAW, bufid)
        call pvmfpack(INTEGER4, task_id, 1, 1, info)
        call pvmfpack(INTEGER4, j, 1, 1, info)
        do 70, i=1, n
         call pvmfpack(REAL8,a(first_row(task_id),i),j,1,info)
 70    continue
        call pvmfsend(inst(no_slave), type, info)
 60    continue
* nb_of_slaves tasks are being performed
* wait for the end of a task
      next_task      = nb_of_slaves+1
      task_processed = 0
*-------------------------------------------------
*    collect results and send work to slaves
```

```
*-------------------------------------------
        call pvmfmkbuf( PVMRAW, bufidR)
*
* receive any message from any slave
 1234   typeR = -1
        call pvmfrecv(-1, typeR, bufidR)
*       get PVM3 tid of slave
        call pvmfbufinfo(bufidR,nbytes,typeR,tid_slave,info)
        call pvmfunpack(INTEGER4,task_id,1,1,info)
        j = first_row(task_id+1) - first_row(task_id)
        call pvmfunpack(REAL8, y(first_row(task_id)), j, 1, info)
        task_processed = task_processed + 1
        call pvmfinitsend( PVMRAW, bufid)
        if (task_processed.GE.nbtasks)  then
*         send end of work
          j=-1
          call pvmfpack(INTEGER4, j, 1, 1, info)
          call pvmfsend(tid_slave, type, info)
        else
          if (next_task.LE.nbtasks)  then
*           send work to no_slave
            task_id   = next_task
            next_task = next_task+1
            j = first_row(task_id+1) - first_row(task_id)
            call pvmfpack(INTEGER4, task_id, 1, 1, info)
            call pvmfpack(INTEGER4, j, 1, 1, info)
            do 75 i=1, n
             call pvmfpack(REAL8,a(first_row(task_id),i),j,1,info)
 75         continue
            call pvmfsend(tid_slave, type, info)
          else
*           send end of work to slave
            j=-1
            call pvmfpack(INTEGER4, j, 1, 1, info)
            call pvmfsend(tid_slave, type, info)
          endif
*         collect other results from slaves
          goto 1234
        endif
*
* Parallel compution is done.
      call pvmfexit(retcode)
      stop
      end
```

Slave's code:

```
      PROGRAM slaveD

* we only show the most important parts of the code.
* (declarations have been suppressed)
*
      type = 0

* Enroll this program in PVM3
      call pvmfmytid(my_id)

* Get the tid of the master's task id .
      call pvmfparent(p_id)
      call pvmfmkbuf( PVMRAW, bufid)
      call pvmfrecv(p_id, type, bufid)
*
* Receive the number of columns and the vector x
      call pvmfunpack(INTEGER4, n, 1, 1, info)
      call pvmfunpack(REAL8   , x, n, 1, info)
*
* receive the task_id, the size of the block and the sub-matrix
 1234 call pvmfrecv(p_id, type, bufid)
      call pvmfunpack(INTEGER4, task_id, 1, 1, info)
*
* test if end of computation
      if (task_id.EQ.-1) goto 500
      call pvmfunpack(INTEGER4, m, 1, 1, info)
      do 10, i=1,n
        call pvmfunpack(REAL8, a(1,i), m, 1, info)
 10   continue
*
*    compute his part of the work
      call dgemv('N',m,n,one,a,lda,x,incx,zero,y,incx)
*
*    send back task_id and result to master
      call pvmfinitsend( PVMRAW, bufidS)
      call pvmfpack(INTEGER4, task_id, 1, 1, info)
      call pvmfpack(REAL8, y, m, 1, info)
      type = 0
      call pvmfsend(p_id, type, info)
      goto 1234
*
* leave Pvm before exiting
 500  call pvmfexit(info)
*
      stop
      end
```

We show in Figure 8 the traces obtained using XPVM during the parallel execution of the dynamic implementation of our illustrative example. The master process is located on computer node *rosanna*.

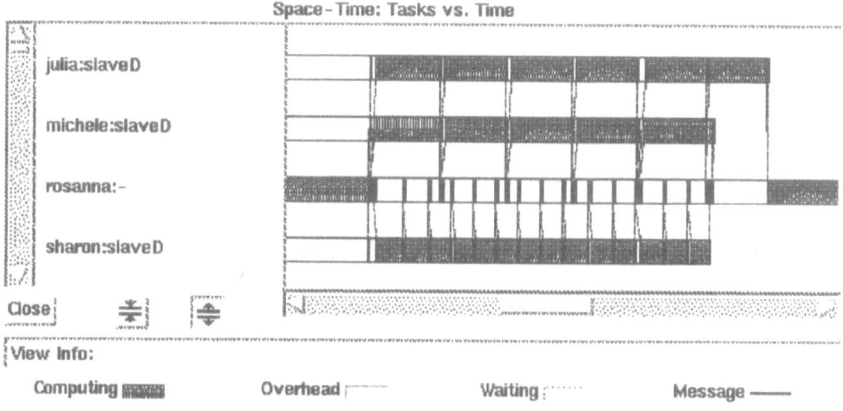

Figure 8. XPVM trace of the dynamic parallelization of the matrix-vector product

We see in Figure 8 that most of the work has been dynamically migrated to the fastest (or less loaded) node computer (*sharon*) of our virtual machine.

4. Examples of industrial codes

We work in a close relationship with industrial partners and have been involved in several studies on porting some of their applications to parallel computers. We discuss in particular two studies performed for Centre National d'Etudes Spatiales (CNES), Toulouse, France (Charles, Daydé, Petitet et al. (1993)), and Aérospatiale, Toulouse, France (Daydé, Duff, L'Excellent, and Giraud (1993)). The aim of these contracts was to evaluate the performance of a selection of application codes on a range of parallel computers: ALLIANT FX/80, ALLIANT FX/2800, BBN TC2000, CRAY-2, CONVEX C220, and networks of workstations. The first part of the studies performed for CNES and Aérospatiale was to evaluate the cost of porting codes to the target computers. The second part concerned the performance obtained by tuning and parallelizing these codes on the target machines. The conclusion of these studies concerned recommendations for a methodology for both porting and developing codes on parallel computers, a performance analysis of the target computers, and comments and recommendations related to the numerical algorithms encountered. Note that the initial parallelization of the applications was always performed on shared memory multiprocessors. We also comment on some performance issues studied within the ESPRIT III EUROPORT-1 'PARALLEL AERO' project.

We will briefly examine in the following subsections the porting of some applications :

- Teledetection and out-of-core solution of linear least-squares problems from CNES,

- Panel method and flutter calculation from Aérospatiale,

- Navier-Stokes multiblock solver from the ESPRIT III EUROPORT-1 'PARALLEL AERO' project.

4.1. TELEDETECTION

The teledetection application consists of a statistical data analysis to identify the main crop zones from satellite images. The image pixels are transformed into a signal using FFT. Then a statistical data analysis is performed on the signals. Since the treatment on each pixel is independent, the code is embarrassingly parallel. Practically, the parallelization is effected at the row level within blocks of 40 image rows that are loaded from the disk at each step.

The total execution time, the CPU time and the time spent in I/O on the BBN TC2000 are reported in Figure 9.

The I/O required for loading part of the image is the main limiting factor for performance on the BBN TC2000. We achieved speed-ups of approximatively 20 on the computational part and 5 on the whole code on 24 processors of the BBN TC2000.

4.2. OUT-OF-CORE SOLUTION OF LINEAR LEAST-SQUARES PROBLEMS

We consider the solution of large full linear least-squares problems. These are solved by using a Cholesky factorization of the normal equations. One use of this software is in the computation of the spherical harmonics of the earth potential from measurements of the gradients of this field. The calculation is carried out in two steps :

- Forming the normal equations

- Cholesky factorization

The normal equations system is full and is partitioned into blocks that are stored on disk. The initial version of the software was entirely written in terms of calls to Level 1 BLAS and LINPACK. Rewriting the software in

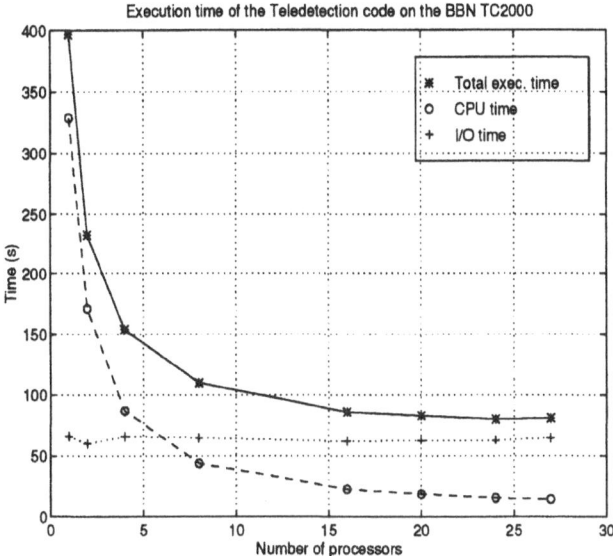

Figure 9. Teledetection application on the BBN TC2000

terms of calls to Level 3 BLAS and to the LAPACK Library (Anderson, Bai, Bischof et al. (1992)) provided a substantial improvement in performance. The parallelization is straightforward because parallel versions of the BLAS are available on some computers. Note that the parallelization in this case is restricted to single blocks. We report in Table 5 the results obtained. The speed-ups achieved are reported in parentheses.

Computer	Procs	Forming $A^t.A$		Cholesky factorization	
		Initial	Tuned	Initial	Tuned
ALLIANT FX/80	1	532	110	259	79
	8	181 (2.9)	28 (3.9)	132 (1.9)	35 (2.3)
CONVEX C220	1	159	103	103	94
CRAY-2	1	25	9	17	8

TABLE 5. Execution time in seconds and speed-up of the initial version (Level 1 BLAS and LINPACK) and of the tuned version (Level 3 BLAS and LAPACK) of the out-of-core linear least-squares solver

In the initial version of code for forming the normal equations, the computation of $A^t.A$ was performed using dot products that represent 47%

of the execution time. The rest of the execution time was spent in I/O. In the Level 3 BLAS version, the matrix-matrix multiplication (GEMM) used instead of the dot products, only requires 0.5% of the execution time. The Cholesky factorization step has the same behaviour. The I/O is therefore the main bottleneck after substitutions of the computational kernels.

4.3. PANEL METHOD

We consider the parallelization of a 3D code (Daydé, Duff, L'Excellent, and Giraud (1993)) that computes flow around an aircraft in a perfect fluid. The solution technique is a panel method using a mesh surface. The surface of the aircraft is divided into panels. Then, the singularities are computed on each panel. The interactions and the contributions of the panels are assembled into a so-called influence matrix. Then the velocity is computed around the aircraft from the vector of singularities. This requires the solution of a nonlinear system of equations using Gauss-Seidel and LU factorization. The computation of the coefficients of the influence matrix is independent. Thus the calculation is easily parallelizable using appropriate data structures. We report in Table 6 the results we obtain on a range of shared memory computers. The speed-ups achieved by the tuned parallel version over the tuned serial version are reported in parentheses. The test problem has 1537 panels for the wings and fuselage and 225 panels for the wake.

| Computer | Procs | Initial Version | Tuned Version | |
		Serial	Serial	Parallel
ALLIANT FX/80	8	8391	6766	2202 (3.1)
CONVEX C220	2	1691	808	580 (1.4)
CRAY-2	4	140	148	77 (1.9)

TABLE 6. Execution time in seconds of initial and parallel versions of panel method

The speed-ups are somewhat disappointing. The calculation of the influence matrix is totally parallelized while the rest of the application cannot be efficiently parallelized (the parallelism is restricted to loop-level within blocks of the matrix that are loaded from disk).

Amdahl's law gives an upper bound to the achievable speed-up :

$$S_p = \frac{p}{(f + (1-f)*p)}$$

where p is the number of processors, f the fraction of code parallelized, and S_p is the theoretical speed-up for p processors. Infinite speed-up (theoretical speed-up using an infinite number of processors) is given by $1/(1-f)$.

Speed-ups	Number of processors		
	2 procs.	4 procs.	8 procs.
Effective speed-up	1.7	2.6	2.7
Theoretical speed-up	1.7	2.5	3.4
Speed-up on INFLUV	2.0	3.7	6.4

TABLE 7. Effective and theoretical speed-ups of panel method on ALLIANT FX/80

Using Amdahl's law, we can compare in Table 7 the effective speed-ups obtained and the theoretical speed-ups predicted by Amdahl's law on the ALLIANT FX/80 on the execution of the whole code. We also report the speed-ups on INFLUV, the procedure that performs the calculation of the influence matrix which is entirely parallelized. Note that the effective speed-up exceeds the theoretical one using 4 processors on the ALLIANT FX/80 because the theoretical estimate does not include any allowance for parallelizing other parts of the code than in INFLUV.

4.4. FLUTTER CALCULATION

We consider a flutter calculation from Aérospatiale. After cleaning up the code and other optimizations at loop-level (using compiler directives and rewriting some loops), we obtained a tuned version that was much more efficient than the initial one. The code is organized into a set of nested loops :

```
Loop on the mach numbers
    Loop on the structural modes
        Loop on the velocity
            Compute frequency and damping
        End loop
    End loop
End loop
```

The two outer loops are parallelizable since the calculations for each mach number and for each mode are independent. We have only parallelized the loop on the modes since often the application is only for one

		Initial version	Tuned version	
Computer	Procs	Serial	Serial	Parallel
ALLIANT FX/80	8	5100	1717	326 (5.3)
BBN TC2000	19	-	2994	202 (14.8)
CONVEX C220	2	1919	216	119 (1.8)
CRAY-2	4	624	45	15 (3.0)
IBM RS/6000-320	8	-	622	93 (6.7)
Cluster RS/6000	3	-	235	89 (2.6)

TABLE 8. Execution time in seconds and speed-up of initial and tuned versions of the flutter calculation application

mach number. Because of the scalability of this application (the degree of parallelism is equal to the number of modes) and the very good data locality (calculation for each mode requires only local data after an initialization step), this application has been successfully parallelized on a range of shared memory and virtual shared memory computers, and on networks of computers (a network of IBM RS/6000-320 connected by Ethernet and a cluster of IBM RS/6000 950, 550, and 530 connected by SOCC) using the PVM message passing package. We report in Table 8 the gains in execution time achieved from tuning and parallelizing the codes. The calculation involves 38 modes. We report in parentheses the speed-ups achieved by the tuned parallel version over the tuned serial version.

We show in Figure 10 the execution times of the tuned version when varying the number of processors from 1 up to 8 on a set of computers.

The speed-ups are very satisfactory. The effect of implementing an appropriate strategy for handling load balancing efficiently is crucial for the performance. Since the solution times for each mode are unbalanced and unpredictable, a dynamic load balancing scheme was implemented. We have simply taken the self-scheduling strategy used to manage loop-level parallelism on shared memory computers, while on the BBN and on clusters of workstations we have used a master/slave paradigm. We show in Figure 11 the solution times required for each mode on the ALLIANT FX/80.

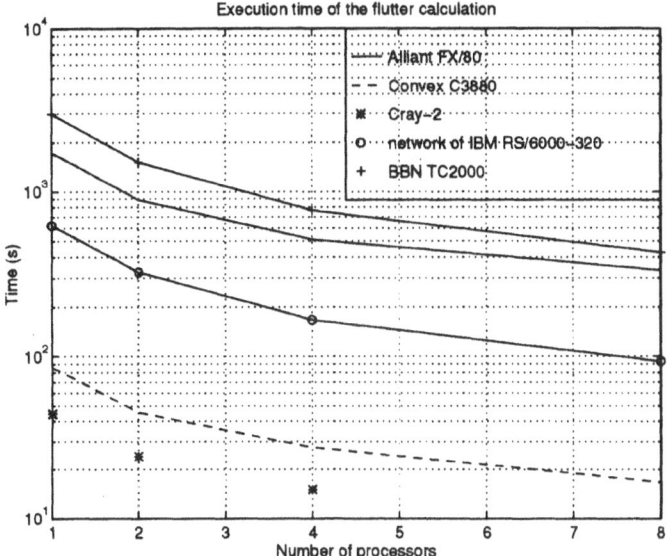

Figure 10. Execution time of the flutter calculation

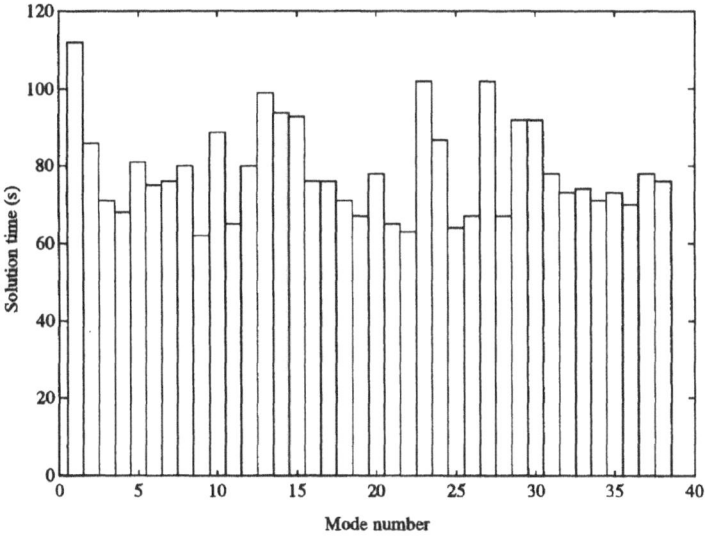

Figure 11. Solution time for each structural mode on ALLIANT FX/80

4.5. ESPRIT III EUROPORT-1 "PARALLEL AERO" PROJECT

We examine some aspects of the "PARALLEL AERO" project. One of the goal (Van Kemenade, Daydé, and Vos (1995)) was to investigate whether the current generation of Massively Parallel Computers (and also network of computers) is capable of solving Navier-Stokes equations for aerodynamic design is a reasonable execution time. Another aspect was to demonstrate that it is possible to design a portable and scalable CFD code across a range of platforms using message passing libraries such as PVM and PARMACS (Hempel, Hope, Keller, and Krotz (1995)).

Our intention in this section is mainly to illustrate how the memory hierarchy and the type of processor (vector or RISC) can affect the code design.

4.5.1. NSMB

The code used in PARALLEL AERO is a Navier-Stokes multiblock solver, called **NSMB**, which is being developed within a joint research project by research centers (IMHEF-EPFL, Lausanne, KTH, Stockholm, and CER-FACS, Toulouse) and two industries (Aérospatiale and SAAB).

NSMB is a multi block solver, and was designed for parallel computations by making calculations within a block independent of other blocks. Synchronization and communication between blocks are required to exchange information at block interfaces, and to control the global convergence. Implementation on distributed memory computers is straightforward, since all data are local to a block. **NSMB** uses a data base, called MEM-COM, to store and retrieve the data for a calculation (coordinates, solution, boundary condition information, ..)

The domain partitioning is a preprocessing step producing a data base which can be directly used by **NSMB**. The domain partitioning tool **MB-Split** gives the mapping of the blocks on the processors, trying to solve the load balancing problem, and gives the list of neighbouring nodes for each processor.

The parallelization of **NSMB** is based on the following concepts :

1. The use of a master/slave paradigm.

2. The master performs all accesses to the MEM-COM data base.

3. The source code is unique for serial or parallel execution

4. The communication implemented in the parallel version of NSMB uses the most portable communication primitives such as send, blocking receive, and non-blocking receive.

The evaluation of the initial **NSMB** uniprocessor performance demonstrated that **NSMB** is well suited for running efficiently on vector processors, but that an efficient implementation on RISC processors would require modifications of the code. We have then studied how to increase the performance of **NSMB** on RISC-processors, while at the same time maintaining the good efficiency on vector processors. Several modifications were included in the industrial version of **NSMB**. We illustrate the method we have used on one of the most time-consuming routines of **NSMB** : **gradvelo**.

4.5.2. *Routine* gradvelo

The routine **gradvelo** computes the velocity gradient at a surface in one direction. For the sake of clarity, we only comment on one of the main computational loop. The velocity gradients are computed into the arrays du, dv, and dw that are declared as follows :

```
    real*8 du(3,-1:n1+2,-1:n2+2,-1:n3+2),
   *       dv(3,-1:n1+2,-1:n2+2,-1:n3+2),
   *       dw(3,-1:n1+2,-1:n2+2,-1:n3+2)
```

In the initial version of **gradvelo**, 1D-loops over all the cells were used to enhance the performance on vector processors, for example :

```
      do 703 n=2*nst+1,nend
c
         l       = n - 11
c
         usurf   = quart*(w(n,2)+ w(n-13,2)+ w(1,2)+ w(1-13,2))
         vsurf   = quart*(w(n,3)+ w(n-13,3)+ w(1,3)+ w(1-13,3))
         wsurf   = quart*(w(n,4)+ w(n-13,4)+ w(1,4)+ w(1-13,4))
c
         sx      = (s(n,1,ns3) + s(1,1,ns3))
         sy      = (s(n,2,ns3) + s(1,2,ns3))
         sz      = (s(n,3,ns3) + s(1,3,ns3))
c
         l       = n - 13
c
         du(n,1) = du(n,1) - usurf * sx
         du(n,2) = du(n,2) - usurf * sy
         du(n,3) = du(n,3) - usurf * sz
c
         du(1,1) = du(1,1) + usurf * sx
         du(1,2) = du(1,2) + usurf * sy
         du(1,3) = du(1,3) + usurf * sz
c
         dv(n,1) = dv(n,1) - vsurf * sx
         dv(n,2) = dv(n,2) - vsurf * sy
         dv(n,3) = dv(n,3) - vsurf * sz
c
         dv(1,1) = dv(1,1) + vsurf * sx
```

```
              dv(1,2) = dv(1,2) + vsurf * sy
              dv(1,3) = dv(1,3) + vsurf * sz
c
              dw(n,1) = dw(n,1) - wsurf * sx
              dw(n,2) = dw(n,2) - wsurf * sy
              dw(n,3) = dw(n,3) - wsurf * sz
c
              dw(1,1) = dw(1,1) + wsurf * sx
              dw(1,2) = dw(1,2) + wsurf * sy
              dw(1,3) = dw(1,3) + wsurf * sz
c
     703    continue
```

As it can be seen in the code below, the access to arrays du, dv, and dw is far to be optimal since, for example, $du(n,1)$, $du(n,2)$, and $du(n,3)$ are referenced within the innermost loops. The stride between these elements is large (equal to the leading dimension of du) and may be critical in some cases.

When executing do-loops, some strides are called *critical* because they generate a very high cache miss ratio (for example when referencing cache lines that are mapped into the same physical location of the cache). These critical strides obviously depend on the cache management strategy. For example, in the execution of the following loop :

```
              do i=1,n,4
                  temp = temp + a(i)
              enddo
```

each read of a(i) will cause a cache miss, assuming that a(i) is one word and that the cache line length is equal to four words (assuming that the cache is initially empty).

The elimination of such strides greater than one in some routines of NSMB is crucial for improving the performance on computers that possess a memory hierarchy involving a cache. The large strides in innermost computational loops are suppressed by permuting the first and the last dimensions of some arrays. Accessing $du(1,n)$, $du(2,n)$, and $du(3,n)$ is obviously more efficient, since these elements are located in the same or in consecutive cache lines. The possibility of reducing the amount of calculations in ghost cells by using three nested loops instead of a long 1D-loop over all the cells was also implemented.

We have used three nested loops in the ijk version and whe have additionally permuted the first and the last dimension of the array du, dv, and dw in the $ijk - flipped$ version. In the $ijk - flipped$ version, the declaration of the arrays becomes :

```
     real*8 du(3,-1:n1+2,-1:n2+2,-1:n3+2),
     *       dv(3,-1:n1+2,-1:n2+2,-1:n3+2),
```

```
*         dw(3,-1:n1+2,-1:n2+2,-1:n3+2)
```

while the calculations are expressed as :

```
      do k=1,n3+2-i2d
        do j=1,n2+2
          do i=1,n1+2
c
            il = i-il1
            jl = j-jl1
            kl = k-kl1
c
            usurf = quart *
*               (w(i  , j, k,2)+w( i-il3, j-jl3, k-kl3,2)
*               +w(il,jl,kl,2)+w(il-il3,jl-jl3,kl-kl3,2))
            vsurf = quart *
*               (w(i  , j, k,3)+w( i-il3, j-jl3, k-kl3,3)
*               +w(il,jl,kl,3)+w(il-il3,jl-jl3,kl-kl3,3))
            wsurf = quart *
*               (w(i  , j, k,4)+w( i-il3, j-jl3, k-kl3,4)
*               +w(il,jl,kl,4)+w(il-il3,jl-jl3,kl-kl3,4))
c
            sx = s(i,j,k,1,ns3) + s(il,jl,kl,1,ns3)
            sy = s(i,j,k,2,ns3) + s(il,jl,kl,2,ns3)
            sz = s(i,j,k,3,ns3) + s(il,jl,kl,3,ns3)
c
            il = i-il3
            jl = j-jl3
            kl = k-kl3
c
            du(1,i,j,k) = du(1,i,j,k) - usurf * sx
            du(2,i,j,k) = du(2,i,j,k) - usurf * sy
            du(3,i,j,k) = du(3,i,j,k) - usurf * sz
c
            du(1,il,jl,kl) = du(1,il,jl,kl) + usurf * sx
            du(2,il,jl,kl) = du(2,il,jl,kl) + usurf * sy
            du(3,il,jl,kl) = du(3,il,jl,kl) + usurf * sz
c
            dv(1,i,j,k) = dv(1,i,j,k) - vsurf * sx
            dv(2,i,j,k) = dv(2,i,j,k) - vsurf * sy
            dv(3,i,j,k) = dv(3,i,j,k) - vsurf * sz
c
            dv(1,il,jl,kl) = dv(1,il,jl,kl) + vsurf * sx
            dv(2,il,jl,kl) = dv(2,il,jl,kl) + vsurf * sy
            dv(3,il,jl,kl) = dv(3,il,jl,kl) + vsurf * sz
c
            dw(1,i,j,k) = dw(1,i,j,k) - wsurf * sx
            dw(2,i,j,k) = dw(2,i,j,k) - wsurf * sy
            dw(3,i,j,k) = dw(3,i,j,k) - wsurf * sz
c
            dw(1,il,jl,kl) = dw(1,il,jl,kl) + wsurf * sx
```

```
            dw(2,il,jl,kl) = dw(2,il,jl,kl) + wsurf * sy
            dw(3,il,jl,kl) = dw(3,il,jl,kl) + wsurf * sz
c
        end do
      end do
    end do
```

We have studied the performance of **gradvelo** using the three versions described :

1. the *initial* version of the code using 1D-loop,

2. the *ijk* version using three nested loops,

3. and the *ijk − flipped* version using three nested loops and where first and last dimensions of arrays *du*, *dv*, and *dw* are flipped.

We report on experiments on a range of problem sizes that correspond to the block sizes used in a real test case, the medium grid of the F5-Wing test case partitioned into different number of blocks :

1. 64 blocks : 12x10x6

2. 32 blocks : 12x10x12

3. 16 blocks : 12x20x12

4. 8 blocks : 24x20x12

5. 4 blocks : 24x20x24

6. 2 blocks : 48x20x24

7. 1 block : 96x20x24.

The average performance improvement achieved over all the block sizes using the *ijk* and *ijk − flipped* versions with respect to the *initial* version is reported in Table 9.

The *ijk* version provides on some machines a slight performance improvement due to the reduction in the number of floating point operations. The gains when using the *ijk − flipped* version are very impressive (except on HP), especially on the IBM 750 and SP2. The improvement factor is between 1.0 and 2.3 depending on the computer.

The cache organization has a strong impact on the performance improvement. The HP has a direct-mapped cache and in this case there is no performance improvement while the performance improvement is very high on computers having a 4 way set-associative cache such as the IBM Power.

We noticed that the performance of the original version of **NSMB** was falling dramatically on some computers for certain block sizes (block sizes

Computer	ijk	$ijk - flipped$
Meiko CS2-HA	1.1	1.7
DEC Alpha	1.2	1.6
IBM 750	1.1	2.3
IBM SP2	1.1	2.3
HP 715/64	0.8	1.0
INTEL PARAGON	1.0	1.4
SUN 10/41	1.0	1.5
SUN Sparc20/50	1.2	2.1

TABLE 9. Average gain in execution time of ijk and $ijk - flipped$ versions vs initial version of **gradvelo**.

equal to 12x10x12, 12x20x12, and 24x20x12). On the IBM 750 and on the IBM SP2, the block size 12x20x12 corresponding to the 16 block decomposition of the medium grid of the F5-wing, had a very bad performance. This was due to an array length for du, dv, and dw equal to (12+4)*(20+4)*(12+4)=6144, i.e. a multiple of the number of cache lines.

The Figure 12 demonstrates that the $ijk - flipped$ version solves the problem of critical strides on IBM SP2. This was also observed on other computers.

The same approach was used in the most time-consuming routines of **NSMB**. Both the *initial* (using $1D$ loops) and the ijk versions of these routines are included in the industrial version of **NSMB** and selected according to the processor used. The *flipped* versions of these routines are automatically generated when requested. Other optimizations (loop unrolling and improving communications) have also been included in the $ijk - flipped$ version of **NSMB**.

We report in Figure 13 the execution time of the *initial* and the $ijk - flipped$ versions of NSMB on the T3D. The benchmark test is the turbulent flow around a 3D wing inside a wind-tunnel in the transsonic regime (F5-Wing test). The grid size is 96x24x20 cells (medium grid), we perform 100 iterations and there is one block per processor. The number of flops used for computing the MFlop rate is that of the 1 block execution, i.e. we do not take into account the arithmetic overhead due to the splitting of a block into smaller blocks. Splitting into smaller blocks introduces a large overhead :

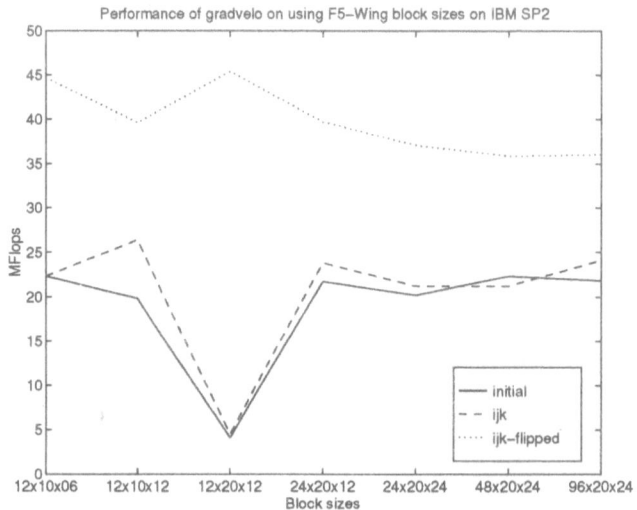

Figure 12. Performance of **gradvelo** on IBM SP2

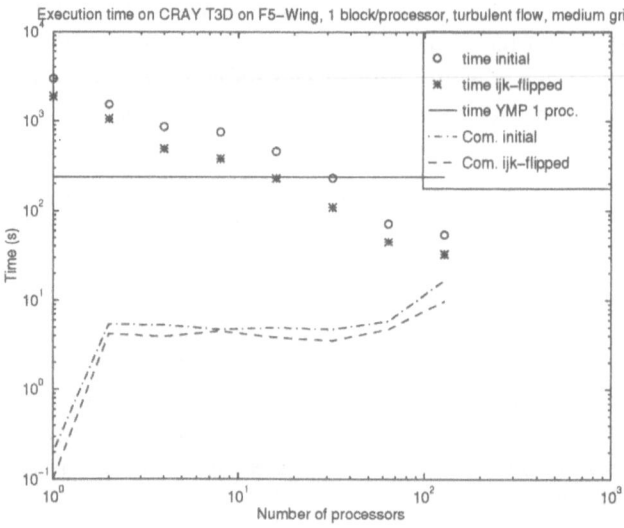

Figure 13. Execution time of **NSMB** on CRAY T3D, F5-Wing, turbulent flow, medium grid, 1 block/proc.

the execution time on the CRAY YMP is almost doubled when using 16 blocks instead of 1. We also include the execution time using 1 block on 1 processor of the CRAY YMP and the communication time for the *initial* and the $ijk - flipped$ versions. In Figure 14 we report the performance and Figure 15 we report the speedup of **NSMB**.

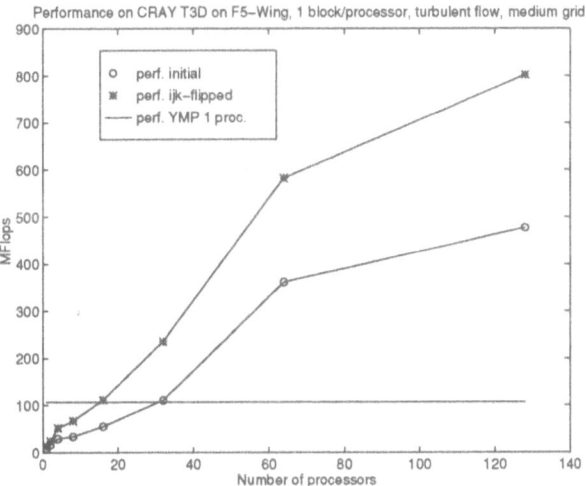

Figure 14. Performance of **NSMB** on CRAY T3D, F5-Wing, turbulent flow, medium grid, 1 block/proc.

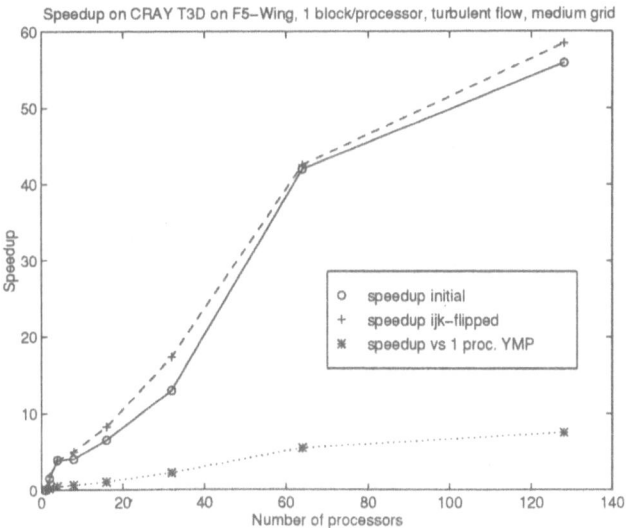

Figure 15. Speedup of **NSMB** on CRAY T3D, F5-Wing, turbulent flow, medium grid, 1 block/proc.

5. Conclusion

There are some industrial codes that are good candidates for easy parallelization. Some other codes require a change of solution techniques or a reorganization of the calculations. Nevertheless, we have seen that the gains achieved do not always come from parallelization but may come from cleaning up codes or from vectorization.

We have shown how the architecture of a computer – type of processor : either RISC or vector, and the memory organization – may influence the code design.

Most of the complications that arise when porting industrial codes to parallel computers come from the lack of maturity in programming environments (languages, debuggers, performance analyzers,..) and algorithm design (since most of the codes have been developed on serial machines).

Virtual shared memory is a very convenient feature that simplifies the porting of existing codes. It also demonstrates how shared memory can scale to hundreds of processors. But data locality is still crucial for efficiency. Therefore, most of the tuning aims at finding a good data distribution which is exactly the same problem as on multicomputers using message passing programming model.

Massively parallel computers are increasingly considered as an alternative to conventional high performance computers. They compare very favourably in peak performance with shared memory computers of a similar cost. However, the application performance depends to a large extent on the ratio of the bandwidth to the computing power which tends to be poor compared with shared memory computers. There are still many research issues concerning the effective use of massively parallel computers as a production facility.

The use of heterogeneous networks of computers as a computing resource is quite attractive, because it is cost-effective and because of the availability of programming environments such as PVM that can also be used within a single parallel computer. But, from our experience, it is restricted to applications where the amount of communication is small compared with the computation.

Parallelizing compilers may one day be capable of generating efficient parallel code for multicomputers (and networks of computers) from sequential code but they still have a long way to go.

References

P. R. AMESTOY AND I. S. DUFF, (1989), *Vectorization of a multiprocessor multifrontal code*, Int. J. of Supercomputer Applics., **3**, 41 59.

P. R. AMESTOY AND I. S. DUFF, (1993), *Memory allocation issues in sparse multiprocessor multifrontal methods*, Int. J. of Supercomputer Applics., **7**, 64 82.

P. R. AMESTOY, M. J. DAYDÉ, I. S. DUFF, AND P. MORÈRE, (1995), *Linear algebra calculations on a virtual shared memory computer*, Int Journal of High Speed Computing, **7**, 21 43.

P. R. AMESTOY, (1991), *Factorization of large sparse matrices based on a multifrontal approach in a multiprocessor environment*, phd thesis, Institut National Polytechnique de Toulouse. Available as CERFACS report TH/PA/91/2.

E. ANDERSON, Z. BAI, C. BISCHOF, J. DEMMEL, J. DONGARRA, J. DU CROZ, A. GREENBAUM, S. HAMMARLING, A. MCKENNEY, S. OSTROUCHOV, AND D. SORENSEN, (1992), *LAPACK : A portable linear algebra library for high-performance computers*, SIAM, Philadelphia.

A. BEGUELIN, J. DONGARRA, A. GEIST, AND V. SUNDERAM, (1993), *Visualization and debugging in an heterogeneous environment*, IEEE Comp., **26(6)**, 88 95.

A. BEGUELIN, J. DONGARRA, A. GEIST, R. MANCHEK, AND V. SUNDERAM, (1991), *A users' guide to PVM Parallel Virtual Machine*, Tech. Rep. ORNL/TM-11826, Oak Ridge National Laboratory, Oak Ridge, Tennessee.

A. BEGUELIN, J. DONGARRA, A. GEIST, R. MANCHEK, AND V. SUNDERAM, (1995), *Recent enhancements to PVM*, Int Journal of Supercomputer Applications, **9**, 108 127.

A. BEGUELIN, (1993), *Xab: a tool for monitoring PVM programs*, in Workshop on Heterogeneous Processing, IEEE Computer Society Press, 92 97.

R. BUTLER AND E. LUSK, (1992), *Users' Guide to the P4 Parallel Programming System*, tech. rep., University of North Florida, Argonne National Laboratory.

J. L. CHARLES, M. J. DAYDÉ, A. PETITET, L. PRÉVOST, AND E. SIMMONET, (1993), *Evaluation de calculateurs multiprocesseurs pour les logiciels et bibliothèques scientifiques du CNES: Rapport Final*, tech. rep., CERFACS, Toulouse, France.

J. CHOI, J. DEMMEL, I. DHILLON, J. DONGARRA, S. OSTROUCHOV, A. PETITET, K. STANLEY, D. WALKER, AND R. C. WHALEY, (1995), *ScaLAPACK: A Portable Linear Algebra Library for Distributed Memory Computers - Design Issues and Performance*, Tech. Rep. LAPACK Working Note 95, CS-95-283, University of Tennessee.

J. CHOI, J. DONGARRA, S. OSTROUCHOV, A. PETITET, D. WALKER, AND R. C. WHALEY, (1995), *A Proposal for a Set of Parallel Basic Linear Algebra Subprograms*, Tech. Rep. LAPACK Working Note 100, CS-95-283, University of Tennessee.

M. J. DAYDÉ AND I. S. DUFF, (1991), *Use of Level 3 BLAS in LU factorization in a multiprocessing environment on three vector multiprocessors, the ALLIANT FX/80, the CRAY-2, and the IBM 3090/VF*, Int. J. of Supercomputer Applics., **5**, 92 110.

M. J. DAYDÉ AND I. S. DUFF, (1995), *Porting industrial codes and developing sparse linear solvers on parallel computers*, Computing Systems in Engineering, **6**, 295 305.

M. J. DAYDÉ AND I. S. DUFF, (1996), *A Block Implementation of Level 3 BLAS for RISC Processors*, Tech. Rep. to appear, ENSEEIHT-IRIT.

M. J. DAYDÉ, I. S. DUFF, AND A. PETITET, (1992), *A Parallel Block Implementation of Level 3 BLAS Kernels for MIMD Vector Processors*, Tech. Rep. TR/PA/92/74, CERFACS, Toulouse, France.

M. J. DAYDÉ, I. S. DUFF, J. Y. L'EXCELLENT, AND L. GIRAUD, (1993), *Evaluation d'ordinateurs vectoriels et parallèles sur un jeu de programmes représentatifs des calculs à la division avions de l'Aérospatiale : Rapport Final*, Tech. Rep. FR/PA/93/19, CERFACS, Toulouse, France.

M. J. DAYDÉ, V. VAN KEMENADE, AND J. B. VOS, (1995), *Description, Validation and Evaluation of the New Parallel Version of NSMB*, Tech. Rep. Esprit Project 8421 : PARALLEL AERO, WP6.1, Deliverable RD21.

J. J. DONGARRA AND E. GROSSE, (1987), *Distribution of Mathematical Software Via Electronic Mail*, Comm. ACM, **30**, 403 407.

J. DONGARRA AND R. C. WHALEY, (1995), *A Users' Guide to the BLACS*, Tech. Rep. CS-95-281, University of Tennessee, Knoxville, Tennessee, USA.

J. J. DONGARRA, J. DU CROZ, I. S. DUFF, AND S. HAMMARLING, (1990a), *Algorithm 679. A set of Level 3 Basic Linear Algebra Subprograms.*, ACM Transactions on Mathematical Software, **16**, 1 17.

J. J. DONGARRA, J. DU CROZ, I. S. DUFF, AND S. HAMMARLING, (1990b), *Algorithm 679. A Set of Level 3 Basic Linear Algebra Subprograms: model implementation and test programs*, ACM Transactions on Mathematical Software, **16**, 18 28.

J. J. DONGARRA, J. J. DU CROZ, S. HAMMARLING, AND R. J. HANSON, (1988), *An extended set of Fortran Basic Linear Algebra Subprograms*, ACM Trans. Math. Softw., **14**, 17 and 18 32.

J. DONGARRA, R. HEMPEL, A. J. G. HEY, AND D. W. WALKER, (1995), *MPI : A Message Passing Interface Standard*, Int Journal of Supercomputer Applications, **8**, (3/4).

J. J. DONGARRA, (1992), *Performance of Various Computers Using Standard Linear Algebra Software*, Tech. Rep. CS-89-85, University of Tennessee, Knoxville, Tennessee, USA.

I. S. DUFF AND J. K. REID, (1983), *The multifrontal solution of indefinite sparse symmetric linear systems*, ACM Transactions on Mathematical Software, **9**, 302 325.

I. S. DUFF AND J. K. REID, (1984), *The multifrontal solution of unsymmetric sets of linear systems*, SIAM Journal on Scientific and Statistical Computing, **5**, 633 641.

I. S. DUFF, R. G. GRIMES, AND J. G. LEWIS, (1992), *Users' Guide for the Harwell-Boeing Sparse Matrix Collection (Release I)*, Technical Report RAL 92-086, Rutherford Appleton Laboratory.

FORTRAN 90, (1991), *ISO/IEC 1539:1991 (E) and now ANSI X3.198-1992*, tech. rep.

A. GEIST, A. BEGUELIN, J. DONGARRA, W. JIANG, R. MANCHEK, AND V. SUNDERAM, (1993), *PVM 3 User's Guide and Reference Manual*, Tech. Rep. ORNL/TM-12187, Engineering Physics and Mathematics Division, Oak Ridge National Laboratory, Tennessee.

S. HARIRI AND A. VARMA, (1993), *High-Performance Distributed Computing : Promises and Challenges*, Concurrency : Practice and Experience, **5**, 233 238.

R. HEMPEL, H.-C. HOPE, U. KELLER, AND W. KROTZ, (1995), *PARMACS V6.1 Specification*, Tech. Report GmbH Technical Report, PALLAS.

HIGH PERFORMANCE FORTRAN FORUM, (1993), *High Performance Fortran Language Specification*, tech. rep., Rice University, Houston, Texas.

C. L. LAWSON, R. J. HANSON, D. R. KINCAID, AND F. T. KROGH, (1979a), *Basic Linear Algebra Subprograms for Fortran Usage*, ACM Transactions on Mathematical Software, **5**, 308 323.

C. L. LAWSON, R. J. HANSON, D. R. KINCAID, AND F. T. KROGH, (1979b), *Algorithm 539. Basic linear algebra subprograms for Fortran usage*, ACM Trans. Math. Softw., **5**, 324 325.

R. SCHREIBER AND H. D. SIMON, (1992), *Towards the Teraflop in CFD*, tech. rep., NASA Ames Research Center, Moffett Field, CA.

V. VAN KEMENADE, M. J. DAYDÉ, AND J. B. VOS, (1995), *Parallel Navier-Stokes Multi-Block Code to Solve Industrial Aerodynamic Design Problems on High Performance Computers*, in Proceedings of HPCN 95 Europe, Milano, Italy.

LOAD BALANCING FOR COMPUTATIONAL FLUID DYNAMICS CALCULATIONS

An introduction

M. STRENG
Department of Applied Mathematics
University of Twente
P.O. box 217
7500 AE Enschede
The Netherlands
e-mail: m.streng@math.utwente.nl

1. Introduction

Many problems in computational fluid dynamics can be solved numerically by representing the flow on a finite set of points, the *grid*. The solution procedure typically proceeds by repeated application of some calculations for each point x in the grid, using the values of the flow-field in several grid points in a neighbourhood of x. Provided this neighbourhood is not too large, a common way to perform these calculations on a parallel computer, is by employing domain-decomposition. In such a technique, the computational domain is partitioned into subdomains. Each subdomain contains a subset of all the points in the original grid, and is assigned to a processor of the parallel machine. Each processor performs the calculations for the points in the subdomain(s) assigned to it. For the computations for the points on the boundaries between two subdomains, information is needed concerning the flow-field in some points in the neighbouring subdomains. Therefore, the processors have to communicate at regular intervals. E.g. in the solution of a time-dependent flow-problem, communication will typically take place (at least) every timestep. The parallel computer is used optimally if none of the processors have to wait for information they need from other processors. Therefore it is important to assign subdomains representing appropriate amounts of work to each processor.

The exact meaning of the word "appropriate" depends on the computational problem and the parallel computer. As an illustrating example, we

P. Wesseling (ed.), High Performance Computing in Fluid Dynamics, 145–172.
© 1996 *Kluwer Academic Publishers.*

will consider the case that for each grid point the same floating-point operations have to be performed. If all the processors can do these operations equally fast (i.e. we have a *homogeneous* machine) and can be used fully for our flow-problem (i.e. we have a *dedicated* machine), then typically the phrase "appropriate amounts of work" refers to an equal number of grid points per subdomain. If the computer has processors which are not all equally fast, then the sizes of the subdomains depend of the speed of the processors: faster processors should deal with more grid points than slower processors. The determination of a partition into subdomains such that no processor has to wait for another is called load balancing. If this partition is used throughout the whole calculation, we say that the balancing is *static*.

A balanced subdivision for the solution of some problem on a parallel computer depends on both the numerical method to solve the problem, and the parallel machine. If the characteristics of one of these changes in the course of the calculations, an initially balanced subdivision can become unbalanced. The characteristics of a numerical method can change e.g. due to grid-refinement, or due to the fact that an iterative method used to solve a subproblem on a subdomain needs more (or less) iterations than in the beginning of the calculations. The characteristics of the parallel machine can change in the case of a cluster of workstations which is in use by other users as well (i.e. operates in a *multi-user* mode). If an unbalanced partition arises during the computations, one may decide to adapt the subdivision in order to obtain a balanced partition again. In that case we speak of *dynamic* load balancing.

Dynamic load balancing is considerably more complex than static balancing. The most important reason is that (much) CPU-time is consumed by the calculation of the new partition and the reinitialisation of the flow-computation, which usually involves transfer of some gridpoints between processors. In general it is not known in advance for how long the new partition will remain balanced. Therefore one has to estimate whether the loss in CPU-time due to proceeding with an unbalanced distribution will exceed the CPU-time needed to repartition the domain and reinitializing the flow-computation. Another difficulty arises because in order to reduce the costs of a rebalancing, one tries to find a new partition which resembles the old one. The application of some of the most efficient static balancing methods may then be cumbersome, because a partitioning obtained with these methods can change dramatically if the grid is changed only a little. Therefore other methods have been developed. However, the static distribution is usually constructed according to some optimality criterion which might not be satisfied again by the rebalanced partition.

1.1. OVERVIEW

Since this text is meant to be of introductory character, we make some simplifications. One important simplification is that we we do not consider the question which subdomains are assigned to which processor. We will assume instead that we have as many subdomains as there are processors, and that no two subdomains are assigned to the same processor. Moreover, we assume that the communications network between the processors is homogeneous in the sense that the cost of transferring a message between two processors is equal for any pair of processors.

We will adopt a restriction which greatly simplifies the balancing problem, namely that for any given partition of the flow-domain, the computational work for each subdomain is known in advance. This is for example the case if the flow-problem is solved using finite differences and with some explicit time-integration method. A case where this assumption is not valid is the case where each sub-problem is solved by some iterative method, where the number of iterations to obtain convergence depends on the details of the flow in that subdomain. Moreover, we assume that the computational work is the same for each grid point. This will in many cases, although not strictly satisfied, be a good approximation.

We introduce some notation and terminology. Throughout the rest of this paper, the letter P is used to denote the number of available processors, and N the number of points in a grid. We assume that the computations for each subdomain proceed by repeatedly executing a computational phase followed by a communication phase, where information concerning points on the boundaries of the subdomains is exchanged. We will say that after execution of a computational phase, the calculations for each grid point have been advanced one *timelevel*.

With the computational grid we can associate an interdependency-graph. The vertices are the grid points, and two vertices are connected by an edge if the numerical computations for one of them cannot be advanced one timelevel if the values of the numerical quantities on the other are not available. In fact, it is this graph which contains all relevant information necessary for the numerical procedure. In the sequel, we will also use the word grid for this graph. If, after some subdivision of the grid, an edge connects two vertices belonging to different subgrids, we say that the edge is *cut* by the partition. The number C is defined as the number of cut edges in some partition. It models the communication volume associated with the partition.

In Section 2 we discuss some well-established methods for static load balancing. We will deal with coordinate and inertial bisection, which use the geometric position of the grid points, and the spectral bisection method,

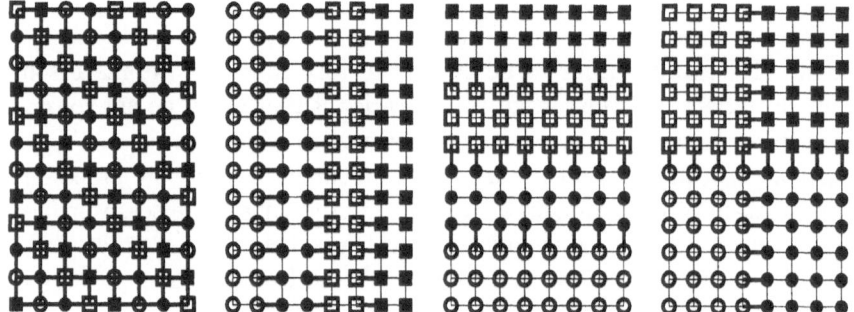

Figure 1.   Several decompositions of a rectangular structured grid in 4 subgrids. The number of cut edges is 172, 48, 32 and 20 respectively.

which is based on information concerning the connectivity of the points in the grid. We will only briefly mention some of the other important approaches, which are not based on bisection. In Section 3 we treat some methods for dynamic load balancing. We will discuss methods based on the bisection methods of Section 2, and two iterative approaches, namely the diffusion method and the dimension exchange method. Finally, we wish to remark that other tutorials on load balancing are already available. A recent one on static balancing is [10].

2. Static load balancing

In the following subsections, we will discuss several methods for static load balancing, starting from a very simple one, and gradually come to the more efficient and elaborate spectral bisection method. First all methods are presented under the assumption that the parallel machine is homogeneous. We will indicate later how to modify them for the heterogeneous case. We will begin with an introductory example.

2.1. INTRODUCTORY EXAMPLE

Consider a 2-dimensional rectangular structured grid, containing $m \times n$ points. We want to divide this grid into P subgrids, each containing an equal number of points. Without further specification, many solutions exist, as indicated in Figure 1 for the case $P = 4$. However, as will be immediately clear, the solutions of Figure 1 are increasingly better from left to right. The reason is that we did not at all incorporate the cost of communication between the subdomains. Intuitively, it will be clear that we have to allocate contiguous blocks of grid points on each processor. The simplest method which does so is the coordinate bisection method. In this method, and in

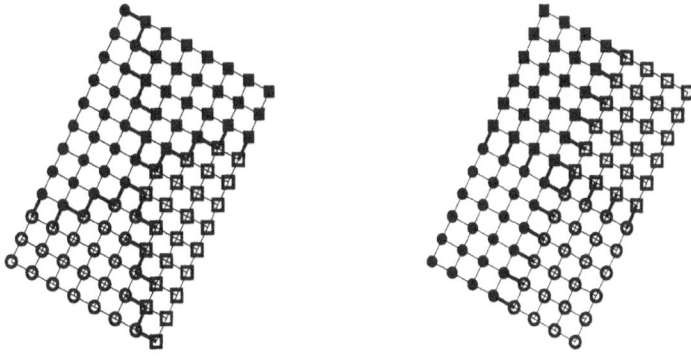

Figure 2.   Two partitions of a rectangular uniform grid, left with recursive coordinate bisection (29 cut edges) and right with recursive inertial bisection (20 cut edges).

the other bisection methods to be presented below, to all grid points a real number σ is assigned according to some rule. The N-vector σ containing these numbers is called the *separator field* for the grid. The points are then ordered with respect to this separator field, and the grid is bisected: the points which have σ_i less than the median value of all the σ_i are grouped into one subset, and the other points into another. The bisection procedure is then applied recursively to these subsets. The term separator field has first been coined by Williams [38].

2.2.   COORDINATE BISECTION

In coordinate bisection methods the grid points are ordered with respect to their x, y or z coordinate. For the example in Figure 1, this results in strip-wise partitioning of the domain, as is shown in the second and third picture in this figure. In general, coordinate bisection produces subdomains with long interfaces, so they lead to large communication volumes. This can be partly overcome by recursive application of alternately x, y (and in 3-d, z) bisection. The grid is first divided into 2 subgrids using e.g. x-bisection. Then to each of the resulting subdomains, y-bisection is applied, etc. This leads to the right-most subdivision of Figure 1. It is clear that this has a much smaller communication volume than can be obtained by x-bisection alone. However, not all grids are partitioned optimally. Consider e.g. the grid of Figure 2. The reason is that the directions for the cuts are specified in advance, so they are in general not related to the actual geometry of the grid. The inertial bisection method often yields much better results.

2.3. INERTIAL BISECTION

Inertial bisection is based on an analogy in mechanics, namely that a solid
body rotates most easily around its principal axis of inertia. This axis is
most likely to be a good approximation to the direction in which the body is
most elongated. In the inertial bisection method, a principal axis of inertia
is associated with the grid. The grid is then cut in a direction perpendicular
to that axis. Let (x_i, y_i, z_i) be the coordinates of the points in the grid, and
define the center of mass (x_m, y_m, z_m) as

$$x_m = \frac{1}{N}\sum_{i=1}^{N} x_i, \quad y_m = \frac{1}{N}\sum_{i=1}^{N} y_i, \quad z_m = \frac{1}{N}\sum_{i=1}^{N} z_i.$$

Further, let I be the 3×3 matrix

$$\begin{pmatrix} I_{xx} & I_{xy} & I_{xz} \\ I_{xy} & I_{yy} & I_{yz} \\ I_{xz} & I_{yz} & I_{zz} \end{pmatrix},$$

where

$$I_{xx} = \sum_{i=1}^{N}(y_i - y_m)^2 + (z_i - z_m)^2,$$

$$I_{yy} = \sum_{i=1}^{N}(x_i - x_m)^2 + (z_i - z_m)^2,$$

$$I_{zz} = \sum_{i=1}^{N}(x_i - x_m)^2 + (y_i - y_m)^2,$$

$$I_{xy} = -\sum_{i=1}^{N}(x_i - x_m)(y_i - y_m),$$

$$I_{xz} = -\sum_{i=1}^{N}(x_i - x_m)(z_i - z_m),$$

$$I_{yz} = -\sum_{i=1}^{N}(y_i - y_m)(z_i - z_m).$$

The eigenvector of I corresponding to the smallest eigenvalue is the principal axis of inertia. The points in the grid are ordered with respect to their
projections on this axis, and the grid is cut into 2 subgrids. The method can
be applied recursively to the subgrids. This will generally provide better
results than the coordinate bisection methods (cf. e.g. [13, 14, 23]). However, even this method not always yields optimal partitions. Consider e.g.

Figure 3. Three partitions of a uniform rectangular grid. Shown are the results obtained from recursive coordinate bisection (30 cut edges), recursive inertial bisection (24 cut edges) and recursive spectral bisection (20 cut edges). It is the same grid as in Figure 2, but now slightly squeezed.

the grid depicted in Figure 3. Although inertial bisection yields a partition with a smaller interface than coordinate bisection, it is evident that the optimal bisection of this grid is given in the last picture of Figure 3.

The reason is that the spatial coordinates of the gridpoints are not always good indicators for the connectivity of the grid. The spectral bisection algorithm ([25, 27, 38]) takes this connectivity explicitly into account.

2.4. SPECTRAL BISECTION

We want to cut the grid G into two parts G_1 and G_2, such that each part contains an equal number of points and the total number of cut edges is minimal. We label the points in the two parts by $+1$ and -1 respectively, and represent the labelling of the points by a vector ξ. An edge connecting the points i and j is cut by the bisection if $\xi_i \xi_j = -1$. Let D be a $N \times N$ diagonal matrix with entries d_{ii} equal to the total number of edges incident with a point i. Further, let A be the $N \times N$ incidence matrix of the graph, with $a_{ij} = 1$ if point i is connected with point j $(i \neq j)$ and 0 otherwise. The number of edges incident with point i which are cut by the bisection is

$$\frac{1}{2}\left(d_{ii} - \sum a_{ij}\xi_i\xi_j\right) = \frac{1}{2}\left(d_{ii}\xi_i\xi_i - \sum_j a_{ij}\xi_i\xi_j\right).$$

The total number of cut edges is then

$$\frac{1}{4}\sum_i \left(d_{ii}\xi_i\xi_i - \sum_j a_{ij}\xi_i\xi_j\right),$$

which can be written, introducing the *Laplacian* matrix $L = D - A$, as

$$\frac{1}{4}\xi^T L \xi,$$

where ξ^T denotes the transpose of ξ. The constraint that both subgrids G_1 and G_2 contain an equal number of points can be cast into the form

$$\mathbf{1}^T \xi = 0,$$

where $\mathbf{1}$ denotes the vector with all entries equal to 1. So we end up with the following optimization problem:

$$\begin{aligned} \text{minimize} \quad & \tfrac{1}{4}\xi^T L \xi, \\ \text{subject to} \quad & \mathbf{1}^T \xi = 0, \\ & \xi_i \in \{-1, +1\}, \quad i = 1, \ldots, N. \end{aligned} \tag{1}$$

Unfortunately, due to the discrete constraint $\xi_i \in \{-1, +1\}$, this problem is hard to solve. Therefore a common heuristic is to relax this constraint, and to replace it by the weaker constraint $\xi^T \xi = N$, which is automatically satisfied by minimizers of (1). The resulting minimization problem can be solved by introducing two lagrange multipliers μ_1, μ_2. A necessary condition for $\bar{\xi}$ to be a local minimizer is

$$\frac{1}{2}L\xi + \mu_1 \mathbf{1} + 2\mu_2 \xi = 0.$$

Since $\mathbf{1}^T L = 0$, left multiplication of this equation by $\mathbf{1}^T$ reveals $\mu_1 = 0$. Then it follows that ξ is an eigenvector of L, corresponding to some eigenvalue λ. The value of the objective function is

$$\frac{1}{4}\xi^T L \xi = \frac{\lambda N}{4}.$$

The matrix L is positive semi-definite, and if the grid G is connected, $\mathbf{1}$ is the only eigenvector corresponding to $\lambda = 0$. This implies that ξ must be chosen as the eigenvector corresponding to the second smallest eigenvalue of L. This vector is usually called the Fiedler vector (cf. [16, 17]). The solution to the original problem (1) is now assumed to be reasonably approximated if one calculates the median of ξ and assigns the value -1 to the points for which ξ_i is below the median, and the value $+1$ to the other points.

The result of applying the spectral bisection algorithm to the graph in Figure 3 yields the optimal partition of the right-most picture of Figure 3. The calculation of the Fiedler vector can be done by a Lanczos method (cf. [18]). However, there exist more efficient (multilevel) methods, cf. [2, 33]. Notice that it may happen that the second smallest eigenvalue has multiplicity larger than 1. This occurs e.g. in the case of a square grid. An eigensolver will in general find two arbitrary eigenvectors corresponding to this eigenvalue. One may then choose some linear combination of these, but, although the choice does not influence the value of the objective in the continuous optimization problem, the procedure for obtaining the discrete vector may result in a bad approximation for the minimal value of the discrete objective function. Finally, the method can easily be extended to incorporate the cost of communication between two processors. Then to each edge is attributed a weight representing this cost, and the matrix A has entries a_{ij} equal to the weight of the edge connecting the points i and j.

2.5. HETEROGENEOUS PARALLEL MACHINES

We briefly indicate how the methods mentioned above can be extended for heterogeneous parallel machines. A typical example of such a machine is a workstation-cluster. Suppose we have P processors, each with a computing speed α_p, $p = 1,\ldots,P$ (measured e.g. by the number of floating-point operations per second). We divide this set into 2 groups $\mathcal{P}_1$ and $\mathcal{P}_2$, each with $P/2$ processors. The groups represent a computing power of

$$\sum_{p \in \mathcal{P}_1} \alpha_p \quad \text{and} \quad \sum_{p \in \mathcal{P}_2} \alpha_p,$$

respectively. Denoting the number of grid points in a grid G by $|G|$, a balanced partition has to satisfy

$$\sum_{p \in \mathcal{P}_1} \alpha_p |G_2| = \sum_{p \in \mathcal{P}_2} \alpha_p |G_1|; \quad |G_1| + |G_2| = |G|.$$

It is straightforward to implement these requirements in the coordinate and inertial bisection methods. In the spectral bisection method, the requirement $\mathbf{1}^T \xi = 0$ has to be replaced by $\mathbf{1}^T \xi = |G_1| - |G_2|$. The resulting continuous optimization problem does not lead anymore to an eigenvalue problem, since now the multiplier μ_1 does not vanish. The solution can be found with the technique discussed in subsection 3.6.2.

2.6. REMARKS

This subsection will contain some general remarks on the bisection-methods.

Remark 1 One straightforward way to extend the methods mentioned so far to a number of processors which is not a power of 2 is the following. We will give only the first step, the recursion is then obvious. The set of processors is divided into two subsets, one containing $\lceil P/2 \rceil$ processors, and the other one $P - \lceil P/2 \rceil$, where $\lceil x \rceil$ denotes the smallest integer larger than x. These subsets can be considered as two processors, each with a relative computing power proportional to the number of processors in each subset. Now the methods of Subsection 2.5 can be applied.

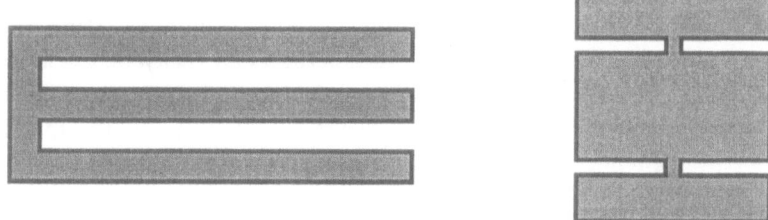

Figure 4. Typical examples of grid-shapes which are split into disconnected components by inertial bisection (left) and spectral bisection (right).

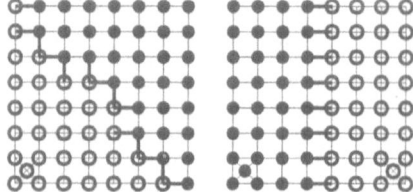

Figure 5. Example of a situation where a slight alteration of the grid can cause the inertial bisection method to produce a completely different partition.

Remark 2 It is a general property of bisection methods that they can lead to subdomains which are not connected. This may be an unwanted situation, which cannot always be avoided. Examples for the inertial and spectral methods are given in Figure 4, and it is possible to construct pathological examples, where one of the subgrids consists of an arbitrary number of connected components. Especially for the spectral bisection method this may (but generally will not) lead to a breakdown of the method in the case that both components contain an equal number of points. Then the eigenvalue 0 has multiplicity 2, and a good cut cannot be found automatically.

Remark 3 Both the inertial bisection method and the spectral bisection method can produce totally different partitions of the grid if the grid is only slightly altered. The reason is that it may happen that the eigenvalue of interest is almost degenerate. See Figure 5 for an example for the inertial bisection method. This makes these methods in principle not directly suitable for dynamic load balancing, although in practice one will not often encounter this phenomenon.

Remark 4 The partition obtained by the spectral bisection method is rather sensitive for the accuracy of the eigenvector-calculation. Especially, inaccurate computation of this eigenvector may cause the number of connected components in the partition to be rather large. Sometimes increasing this accuracy may reveal that the proper subdivision has fewer components. See Figure 6 for an example.

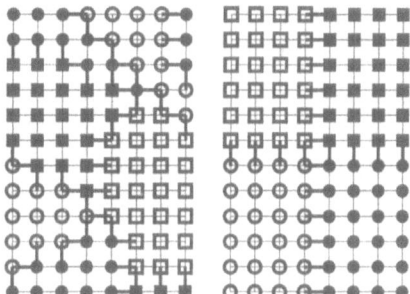

Figure 6. Example where inaccurate eigenvector-computation in the spectral bisection method leads to disconnected components in the partition (left picture). Accurate calculation of this eigenvector would yield the optimal partition on the right.

Remark 5 The choice which of the partitioning methods must be used for a given computational (flow) problem sometimes depends on features not incorporated in these methods. As an example, consider parallel computation of turbulent flow around an airfoil with the Baldwin-Lomax turbulence model [1]. Then it will be convenient if the interfaces of the subdomains are more or less perpendicular to the physical boundaries. As another example, consider the following. All partitioning methods as discussed in this tutorial assume that the computational work is proportional to the number of grid points in the subgrids. This assumption is not always valid. Especially if modern vector- or RISC-processors are used, then the performance may depend strongly on the shape of the subdomains [3, 31].

2.7. OTHER GRID-PARTITIONING METHODS

The techniques discussed so far are all constructive, in the sense that some reasonable quantity is assigned to all grid points, which are then sorted with respect to the value of this quantity. Global properties of the grid are used, and the partitioning is fully deterministic. Besides these techniques, there are many other heuristics. We will only briefly mention a few of them. They differ from the methods discussed before because they only use local properties of the grid (Farhats greedy algorithm) because they are stochastic (simulated annealing) or both (the method of Kernighan-Lin).

The algorithm of Farhat [12], is a greedy algorithm proceeding as follows. Suppose we have already a part (possibly consisting of only one point) of a subdomain. This part is enlarged by adding points connected to points on the boundary of this part. Of course we have to be careful that no points are added which already belong to some subdomain constructed before. The extension of a subdomain can be stopped once this subdomain has the desired size. Therefore, this method is very suitable for obtaining a subdivision for a heterogeneous computer with any number of processors.

An important stochastic approach is based on a model of the turn-around time of an application. This model can include computational aspects, costs of communication, effects of the shapes of the subdomains, hardware characteristics etc. Then the modelled turn-around time is minimized as a function of the grid-partition. This results in a very complex large-scale optimization problem. When this problem is solved using simulated annealing, a partition is constructed in some random way. If this partition results in a lower estimation of the turn-around time, then it is accepted as the current best partition. If it results in a higher estimate, it is accepted with a certain probability. This probability is gradually decreased in the course of the optimization process [35]. Some performance measurements for this method have been made by Williams [38].

A stochastic technique which is often used to enhance a partition obtained by some other method is due to Kernighan and Lin [20]. The idea is that once a grid is partitioned into two subgrids G_1 and G_2, this partition can be improved by exchanging some elements between these subgrids. The method proceeds by performing several sweeps through all the points in G_1 and G_2. At each sweep, two elements are chosen to exchange; if an element has already been involved in an exchange process, it cannot be exchanged another time during that sweep. The partition with the lowest cut-weight found during a sweep is used as starting point for the next sweep. This algorithm gives a locally optimal partition in the sense that simple exchanges of grid points will not reduce the cut weight. There have been some attempts to improve the Kernighan-Lin method. One approach consists of merging nearby vertices, apply the Kernighan-Lin algorithm, undo the merging and re-apply the Kernighan-Lin algorithm [6]. Another approach tries to combine the Kernighan-Lin algorithm with a global optimization method, such as simulated annealing [24].

3. Dynamic load balancing

In the previous section we presented several methods to obtain a decomposition of the computational domain, each constructed with a more or less varying underlying philosophy. It depends on the details of an application which of these methods provides the best result. However, for all methods the aim was to use the parallel machine as efficiently as possible. If in the course of the computations the grid is altered or the characteristics of the machine changes, we have to recompute the partition of the grid. There are several ways to do that. One possibility is to use one of the methods of the previous section to calculate the new partition, allocate the subgrids on the various processors and reinitialize the computations. This has the advantage that the new subdivision is constructed according to the same philosophy as the original, and may so be again well-suited for the particular application. The drawback however is that it may cost much CPU-time to recompute the partition (especially if spectral bisection is used). Moreover, it may cause a large number of gridpoints to be transferred, since the partition may change dramatically if only slight alterations on the grid occur. We have given an example in Remark 3 in the previous section. Another possibility is to use the original partition as a basis, and do only local transfer of gridpoints from one subdomain to adjacent subdomains. The advantage of this method is that the amounts of gridpoints to be transferred can be kept modest. Moreover, the redistribution of grid points can be done completely in parallel. The resulting partition might not fulfill the same optimality criterion as the original one (e.g. it might not have minimal cut-weight).

Because of space limitations, we will not discuss particular applications where dynamic load balancing is used, but merely refer to the following (restrictive) list [30] (Monte Carlo simulations), [7] (heat equation), [21, 22] (Euler/Navier-Stokes equations), [5] (Molecular dynamics), [9] (Finite element methods, general), [37] (Euler equations), [15] (plasma physics). Another topic we will not cover here is the use of dynamic load balancing tools for multi-user cluster computing. An example of such a tool is Load Share Facility (LSF^{TM}) of Platform computing [42]. This is a software package which can be coupled to some message-passing library and tries to distribute all tasks in the cluster evenly over all participating workstations. In fact, it serves as a sort of distributed operating system (cf. also [26]).

As a general remark we wish to say that the dynamic load balancing problem is far from being completely solved. This section should therefore be considered as a summary of some current promising approaches. First, we will give some considerations concerning the performance gain due to dynamic load balancing. Then we present two approaches based on local transfer of gridpoints, the diffusion method and the dimension exchange method, and indicate the relation between them. The method given in Subsection 3.4 elaborates on this relation, and might be very useful for partitions which have been obtained by one of the bisection methods. There are attempts to adapt the spectral bisection method for dynamic load balancing. Some ongoing research is presented in Subsection 3.6. In Subsection 3.7, we briefly consider some problems concerning the transfer of grid points occurring in the adaptation of the partition.

We introduce one more concept. For many purposes, it is convenient to classify the points in a subgrid into level sets. The first level set L_1 contains all points on the boundary of the subdomain. For $q \geq 2$, the level set L_q consists of those points which are connected to a point in level set L_{q-1}, and which are not element of a level set of lower level. Notice that there are only finitely many level sets for each subgrid.

3.1. PERFORMANCE CONSIDERATIONS

In the application of dynamic load balancing to some computational procedure a cycle consisting of the following phases is repeatedly executed:

1. **Solver phase:** Perform some calculations (typically, some steps in a time-integration method) with the given partition of the grid. If necessary, adapt the grid. This phase consists of a repeated application by all processors of a computation step and a communication step.
2. **Statistics phase:** Measure the time each processor needs to process the calculations corresponding to its part of the grid, and how long it has to wait for the results from the neighbouring processors concerning the update of the internal boundary points. This gives an impression of the load-imbalance.
3. **Decision phase:** Decide whether or not to repartition the domain and, if appropriate, decide which method will be used.
4. **Balancing phase:** Calculate the new partition of the grid by application of the method chosen in the previous phase.
5. **Redistribution phase:** Redistribute the domain over the processors and restart the computations. Go to phase 1.

One execution of the phases 1–5 will be called a processing cycle. In a practical situation, the turn-around time of an application can be delayed on the one hand due to a non-optimal distribution of the domain over the processors, and on the other hand due to the costs involved with rebalancing of the domain, i.e. phases 4 and 5. Therefore, the most important phase (but unfortunately also the most difficult one) is the decision phase, where this trade-off must be analyzed. The main difficulty in this phase is that in many cases it is not clear in advance for how long the new partition is the optimal one. Let us state this somewhat more precisely. We assume that the statistics phase and the decision phase cost only a negligible amount of time. If, at processing cycle τ the delay due to load-imbalance is $\Delta_{\text{comp}}^{\tau}$, and the delay caused by redistributing the domain by $\Delta_{\text{bal}}^{\tau}$, then the load balancing algorithm should minimize the total execution time, hence minimize

$$\sum_{\tau} \Delta_{\text{comp}}^{\tau} + \Delta_{\text{bal}}^{\tau}.$$

This is in general very hard to do, because in many cases, after balancing the load at processing cycle τ, an unbalanced domain may arise in processing cycle $\tau+1$. As an example, consider the case where a CFD-problem is solved on a cluster of workstations in a multi-user environment (cf. [11, 29]). Then Δ^τ_{comp} will vary due to the activities of the other users, which cannot be predicted in advance. In a worst case situation, the balancing step may cause a large number of grid points to be transferred at cycle τ, whereas they have to be transferred to the original distribution at cycle $\tau+1$.

One common approach is to do a step towards the partition which has been estimated to be optimal for the solver phase at cycle $\tau+1$. So, instead of transferring all gridpoints in order to achieve the new situation, only a subset of them (typically the ones close to the inter-domain boundaries) are involved in the redistribution, the others are constrained to reside in the same subgrid. This approach can be seen as an iterative method to obtain balanced distributions in the following sense. Suppose the optimal partition of the grid remains identical for all processing cycles $\tau+k$, $k = 1, 2, \ldots$, but is different from that on step τ. Then repeated application of this local transfer of grid points will ultimately approach a globally balanced partition, in the sense that each processor deals with the appropriate amount of points. This is the strategy of the first two methods presented in this section. They might be particularly useful if the load-imbalance is not too large, and hence can be expected to be restored with the transfer of only a modest amount of grid points.

Notice that, although the fluctuations in the grid may be small during the whole computation, it cannot be concluded in general that dynamic load balancing provides only a small performance gain. This is only true in the case that the original partition is a good approximation to *all subsequent* partitions. Consider e.g. the case of P processors, starting each with a grid with P grid points. On the first processor, each processing-cycle P grid points are added. The whole simulation lasts say n processing-cycles. We assume that the processing of 1 grid point costs t seconds CPU-time. If, for simplicity, we ignore the balancing time, the total wall-clock time without load balancing would be $tPn(n+1)/2$ seconds, whereas application of dynamic load balancing would result in a time $(n(n-1)/2+nP)t$ seconds, which, for n large, is smaller by a factor of about the number of processors.

3.2. THE DIFFUSION METHOD

For each $p = 1, \ldots, P$, let X_p^τ (resp. $X_p^{\tau+\frac{1}{2}}$) denote the number of grid-points allocated on processor p after the execution of the solver phase of processing cycle τ (resp. after the execution of the redistribution phase of processing cycle τ). For the moment, we assume that all grid points require an equal amount of work, and that all processors have equal processing speed. We assume that the partition is not balanced. In order to restore the workload on processor p, we will transfer some gridpoints between p and the processors dealing with neighbouring blocks in the flow-domain. The number of gridpoints Δ_{pq} to be transferred from p to a neighbour q is proportional to the load-imbalance between p and q:

$$\Delta_{pq} = w_{pq}(X_p - X_q). \tag{2}$$

We will specify w_{pq} later on, but in any case we will require $w_{pq} = 0$ if no transfer is allowed between p and q (e.g. if p and q are not neighbours). This leads to the new distribution

$$X_p^{\tau+\frac{1}{2}} = X_p^{\tau} - \sum_q w_{pq}(X_p^{\tau} - X_q^{\tau}). \tag{3}$$

We will rewrite this in a simpler form. Consider the *processor interconnection graph* $\mathcal{T}$ with P vertices representing the processors, and edges connecting those vertices corresponding to adjacent subdomains. Each edge has a weight w_{pq}. The $P \times P$ matrix $\mathcal{M}$ is defined as the matrix with entries

$$\mathcal{M}_{pq} = \begin{cases} w_{pq} & p \neq q, \\ 1 - \sum_k w_{pk} & p = q \end{cases} \tag{4}$$

Defining $\mathbf{X}^{\tau}$ as the vector with entries $X_1^{\tau}, \ldots, X_P^{\tau}$, (3) takes the form

$$\mathbf{X}^{\tau+\frac{1}{2}} = \mathcal{M}\mathbf{X}^{\tau}. \tag{5}$$

Notice that $\mathcal{M}$ preserves the total workload: $\sum_p X_p^{\tau+\frac{1}{2}} = \sum_p X_p^{\tau}$.

If all diagonal entries of $\mathcal{M}$ are positive, and the graph obtained from $\mathcal{T}$ by removing all edges having zero weight is connected then the iteration $\mathcal{M}^k \mathbf{X}^{\tau}$ converges for $k \to \infty$ to the balanced partition (cf. [8, 4]). The convergence rate is only linear, and is determined by the second largest eigenvalue of $\mathcal{M}$.

We still have to give reasonable values for the weights w_{pq}. One choice is to take

$$w_{pq} = \frac{1}{1 + \max(d_p, d_q)},$$

where d_p is the number of edges incident to p. For certain regular graphs $\mathcal{T}$, such as the regular mesh and the torus, other choices for w_{pq} can accelerate the convergence (cf. [4]).

If additional work is generated in the solver phase $\tau + 1$, the distribution of the grid points over the processors as time elapses can be written as

$$\mathbf{X}^{\tau+1} = \mathcal{M}\mathbf{X}^{\tau} + \mathbf{W}^{\tau+1},$$

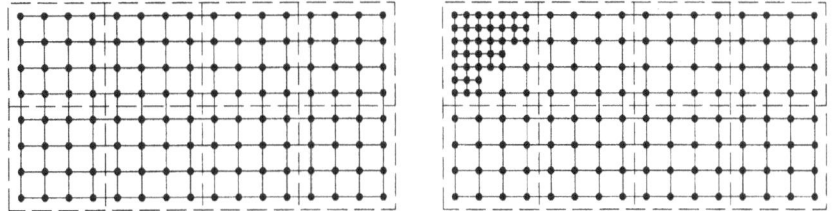

Figure 7. Original grid (left) and refined grid (right) before rebalancing.

where $\mathbf{W}^{\tau+1}$ represents the generation (or deletion) of grid points.

As an example, we consider a rectangular uniform mesh, distributed over 8 processors. Originally, each subdomain contains 16 grid points. At some moment during the computations, the grid in the left-most subgrid is refined, and this subgrid now contains 40 elements. See Figure 7. The interconnection graph $\mathcal{T}$ is shown in Figure 8, as well as the progress of the diffusion method. Since the number of grid points to be transferred is an integer, one must impose a rule what to do if one term $w_{pq}(X_p^\tau - X_q^\tau)$ is not an integer. In the example, these numbers are rounded to the nearest integer. Because of this, the final partition is not completely balanced.

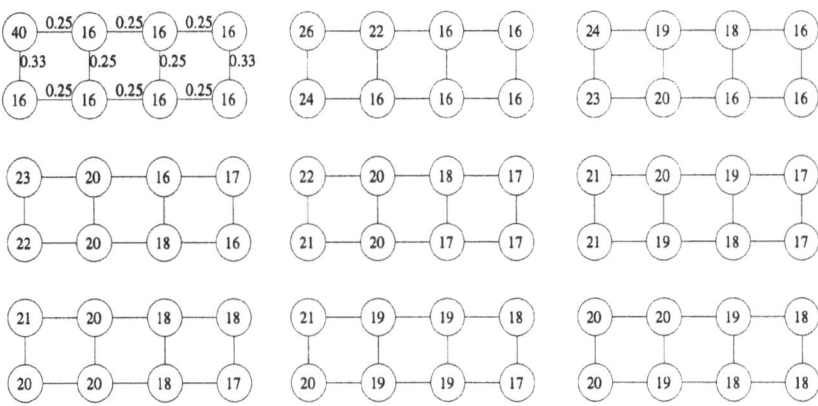

Figure 8. Processor interdependency graph (top left) with the edge-weights and the number of grid points per subdomain. The other pictures indicate the progress of the diffusion method. After 8 iterations the final distribution has been reached.

Notice that the diffusion method can be executed completely in parallel. The difficulty however is that it only gives information about *how many* grid points must be transferred, but not *which* grid points should be involved in the balancing procedure. In many cases, this is a completely non-trivial problem, which is commented upon in subsection 3.7. The convergence of the diffusion method to the globally balanced state is quite slow. In the next subsection we present a method converging (somewhat) faster.

3.3. THE DIMENSION EXCHANGE METHOD

In the dimension exchange method, the edges in the processor interconnection graph $\mathcal{T}$ are coloured, in such a manner that all edges incident with a vertex have different colours. Suppose the maximum number of colours in the graph is c. Then first the load is balanced between processors connected by an edge of colour 1, then between those connected with an edge of colour 2 etc, up to colour c. If two processors p and q balance their load, as in the case of the diffusion method, a number of grid points is transferred which is proportional to the load-imbalance. In step $s = 1, \ldots, c$, if points p and q balance their load, we obtain (put $X_p^{\tau,0} = X_p^\tau$ and $X_p^{\tau+\frac{1}{2}} = X_p^{\tau,c}$):

$$X_p^{\tau,s} = X_p^{\tau,s-1} + \lambda_{pq}(X_q^{\tau,s-1} - X_p^{\tau,s-1}). \tag{6}$$

We can cast this into a simpler form, by defining, for each colour s, $P \times P$ matrices $\mathcal{C}^s$ with entries $C_{pp}^s = 1 - \lambda_{pq}$ if there is a vertex with colour s incident with processor p, $C_{pq}^s = \lambda_{pq}$, if this edge connects processor q with p, and 0 otherwise. Now iteration (6) can be written as

$$\mathbf{X}^{\tau+\frac{1}{2}} = \mathcal{C}^c \cdots \mathcal{C}^2 \mathcal{C}^1 \mathbf{X}^\tau. \tag{7}$$

If $0 < \lambda < 1$, each factor in this matrix-product is a doubly stochastic matrix. Hence the product $C = \mathcal{C}^c \cdots \mathcal{C}^2 \mathcal{C}^1$ has this property as well, and, just as in the case of the diffusion method, the iteration $C^k \mathbf{X}^\tau$ converges for $k \to \infty$ to the globally balanced load distribution. The value of the parameter λ, called the *exchange parameter*, is usually taken as $1/2$. However, for some regular connectivity-graphs $\mathcal{T}$ this is not the optimal choice (cf. [39, 40, 41]). In figure 9, we will repeat the example of the previous subsection, but now using the dimension exchange method. In this case, the final distribution has been reached in three iterations, which is less than required with the diffusion method. The distribution is not completely balanced, due to the fact that integer numbers of grid points must be transferred. This effect is analyzed in more detail by [28].

Just as in the case of the diffusion method, the dimension exchange method can be implemented completely in parallel. Furthermore, also this algorithm does not prescribe *which* grid points have to be transferred.

The advantage of the dimension exchange method over the diffusion method is that convergence is faster, due to the fact that after balancing the load with one neighbour, the new result is used for balancing with the next neighbour. The difference between the methods can be compared with

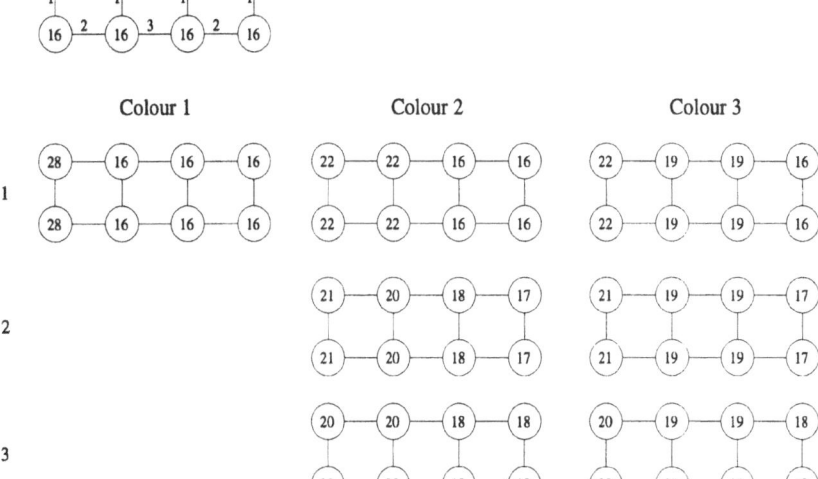

Figure 9. Example of dynamic load balancing with the dimension exchange method. Shown are the colouring of the edges (top left) and the progress of the iteration. In three iterations the final distribution has been reached.

the difference between the Gauss-Seidel and the Jacobi iterative methods to solve linear systems. Also an analogue to multigrid methods has been proposed, which converges still faster, but is harder to implement completely in parallel. However, the load becomes better balanced than by application of the two previously mentioned iterative methods.

3.4. A MULTILEVEL METHOD

In the multilevel method, proposed in [19], the set of processors is divided into two groups, each containing approximately the same number of processors. Each group is regarded as one processor, dealing with one subgrid. Balancing is done between these two groups. The procedure is then applied recursively to each of the groups. The recursion terminates after $\log_2 P$ steps. Denote the number of processors in a set $\mathcal{P}$ by $\|\mathcal{P}\|$. Then, the algorithm runs as follows:

1. set $\mathcal{P} = \{1, \ldots, P\}$
2. bisect $\mathcal{P}$ into subsets $\mathcal{P}_1$ and $\mathcal{P}_2$ such that

   (a) $\mathcal{P}_1 \cup \mathcal{P}_2 = \mathcal{P}$ and $\mathcal{P}_1 \cap \mathcal{P}_2 = \emptyset$,

   (b) $|\|\mathcal{P}_1\| - \|\mathcal{P}_2\|| \leq 1$.

3. Balance the load between $\mathcal{P}_1$ and $\mathcal{P}_2$ by transferring

$$\frac{\sum_{p \in \mathcal{P}_2} X_p^\tau \|\mathcal{P}_1\| - \sum_{p \in \mathcal{P}_1} X_p^\tau \|\mathcal{P}_2\|}{\|\mathcal{P}_1\| + \|\mathcal{P}_2\|}$$

gridpoints from $\mathcal{P}_2$ to $\mathcal{P}_1$.

4. Repeat the procedure for $\mathcal{P}_1$ and $\mathcal{P}_2$:

  - If $\|\mathcal{P}_1\| > 1$, set $\mathcal{P} = \mathcal{P}_1$ and goto step 2.

  - If $\|\mathcal{P}_2\| > 1$, set $\mathcal{P} = \mathcal{P}_2$ and goto step 2.

Notice that the last step can be done in parallel. For the example in the previous subsection the multilevel method gives after one iteration the perfectly balanced partition (all subgrids have 19 grid points). However, in a practical situation, application of this algorithm requires some care. One difficulty in connection with the multilevel method is the transfer of grid points between processor-sets containing more than one processor. The method does not prescribe *which* gridpoints should be transferred, and now this is even more complicated than in the case of the diffusion method and the dimension exchange method, since now these points can be chosen from a set of subdomains instead of from just one. A similar situation occurs for the subdomain(s) to which these points are transferred. Another problem is the appropriate selection of the processor-subsets $\mathcal{P}_1$ and $\mathcal{P}_2$. Choosing them in an arbitrary way may result in grid points being transferred to non-neighbouring processors. A good choice might be to use the hierarchy which is implicitly generated by application of one of the recursive bisection techniques of the previous section. It may seem more natural however, to adapt these bisection methods themselves for dynamic load balancing. We will review some current research concerning the spectral bisection method in Subsection 3.6.

3.5. EXTENSION TO HETEROGENEOUS NON-DEDICATED MACHINES

The dynamic balancing methods mentioned so far can be easily extended to cover the case of heterogeneous computers operating in a multi-user mode. In this case, not the number of grid points in a subgrid is the essential quantity measuring the load-imbalance, but rather the amount of computational work. Suppose we have P processors, each with a computing speed α_p, $p = 1, \ldots, P$ (measured e.g. by the number of floating-point operations per second). If the parallel machine operates in multi-user mode, the computing speed depends on the timelevel τ. We will denote this dependency as α_p^τ. In the iteration-formulas (3) and (6), the quantities X_p^τ should be replaced by $\alpha_p^\tau X_p^\tau$. In the case of the multilevel method, the two groups of processors represent a computing power of

$$A_1^\tau = \sum_{p \in \mathcal{P}_1} \alpha_p^\tau \quad \text{and} \quad A_2^\tau = \sum_{p \in \mathcal{P}_2} \alpha_p^\tau,$$

respectively. The transfer-formula should then be replaced by

$$\frac{\sum_{p \in \mathcal{P}_2} X_p^\tau A_1^\tau - \sum_{p \in \mathcal{P}_1} X_p^\tau A_2^\tau}{A_1^\tau + A_2^\tau}.$$

3.6. MODIFICATIONS OF THE SPECTRAL BISECTION METHOD

If, during the computation, the grid is adapted, and a redistribution of the domain has to be performed, then application of the recursive spectral bisection method might yield a completely different partition. Therefore, modifications are currently being sought which try to overcome this difficulty. We will briefly discuss the idea of some of these.

3.6.1. *Clustering*

In the clustering approach, proposed by [37], the grid points belonging to level-sets with level higher than some number q are clustered and treated as one single vertex. The idea is to use the spectral bisection method to partition the resulting graph, which has considerably fewer vertices than the original graph, so that only points which are close to the boundaries of the subdomains can participate in the transferring process. E.g. if $q = 1$, only boundary-points are possibly moved to other processors. In order to realize this idea, some difficulties must be resolved. The first is that the clustering can lead to vertices which have more than one edge in common with the (new) vertices arising from the clustering. This can be incorporated by collapsing them into one edge with a weight equal to this number of edges. The second difficulty is that some of the entries in the Fiedler vector represent more than one vertex in the original graph. In the calculation of the median, these entries are counted as many times as the number of vertices they represent. It may occur that bisecting the graph with this Fiedler vector, cuts a cluster into two parts, which is not our aim. This will especially happen if a good bisection cannot be obtained by transferring only near-boundary points. If this happens, the process can be started again with a higher value for q, so by clustering fewer points. If the maximal value for q has been reached, then this clustering approach involves the whole graph and is therefore equivalent to full spectral bisection. This may happen if the grid has been changed completely by the adaptation. Of course, if it is known in advance that this will occur, then it is cheaper to apply full spectral bisection from the start. Finally, care must be taken concerning the decision which points to migrate to which processor. We will briefly comment on this topic in Subsection 3.7.

3.6.2. *Incorporating transfer cost*

Attempts are being made to take the costs of transfer of grid points explicitly into account. We will sketch two approaches.

Cost function approach. We will divide the unbalanced graph into two subgraphs, each containing an equal number of vertices, such that the sum of the transfer costs and the communication costs is minimal [32]. Suppose β (resp. ξ) is a vector with entries 1 if the corresponding vertex is in partition 1 of the unbalanced (resp. balanced) graph, and -1 if it is in partition 2. Then the cost of transferring grid points from one partition to the other is modelled by

$$\frac{1}{4}c_t \sum_i (\xi_i - \beta_i)^2,$$

where c_t is the cost to transfer one gridpoint. By an analogous reasoning as used in Subsection 2.4 we obtain the following continuous optimization problem:

$$\begin{aligned}
\text{minimize} \quad & \frac{1}{4}(\xi^T L \xi + c_t \sum_i (\xi_i - \beta_i)^2), \\
\text{subject to} \quad & \mathbf{1}^T \xi = 0, \\
& \xi^T \xi = N.
\end{aligned} \tag{8}$$

Again we introduce two lagrange multipliers μ_1, μ_2. Then a necessary condition for $\bar{\xi}$ to be a local minimizer is

$$\frac{1}{2}(L\bar{\xi} + c_t(\bar{\xi} - \beta)) + \mu_1 \mathbf{1} + 2\mu_2 \bar{\xi} = 0. \tag{9}$$

Since $\mathbf{1}^T L = 0$, left multiplication of this by $\mathbf{1}^T$ reveals $\mu_1 = c_t \mathbf{1}^T \beta / (2N)$. Inserting this into (9) yields, after some rearrangements,

$$L\bar{\xi} = (-4\mu_2 - c_t)\bar{\xi} + c_t(\beta - \frac{\mathbf{1}^T \beta}{N}\mathbf{1}),$$

which can be rewritten as

$$(L - \lambda \mathbf{I}_n)\bar{\xi} = \beta'. \tag{10}$$

This equation has a unique solution for any λ which is not an eigenvalue of L. Notice that $\mathbf{1}^T \beta' = 0$, so solutions of (10) satisfy the constraint $\mathbf{1}^T \bar{\xi} = 0$. Now we assume that the eigenvalues of L are $0 = \lambda_1 < \lambda_2 \leq \ldots \leq \lambda_n$, and the corresponding orthonormal eigenvectors $u_1, \ldots, u_n$. In this basis, L is diagonal, and we write $\beta'_i = b_i u_i$ for the i-th component of β' (note that $\beta'_1 = 0$). Equation (10) can now be solved for $\bar{\xi}$:

$$\bar{\xi} = \sum_{i=2}^{n} \frac{b_i}{\lambda_i - \lambda} u_i.$$

Inserting this solution into the normalization constraint $\bar{\xi}^T \bar{\xi} = N$ yields an equation for λ (and hence for the multiplier μ_2 of the original optimization problem):

$$\sum_{i=2}^{n} \frac{b_i^2}{(\lambda_i - \lambda)^2} = N.$$

This equation may have several solutions, but it is easy to see that there is at least one solution $\bar{\lambda}$, since the left hand side is zero for $\lambda \to -\infty$ and increases monotonically to ∞ for $\lambda \to \lambda_2$. The resulting vector $\bar{\xi}$ approaches the eigenvector of L corresponding to λ_2 if c_t approaches zero and is used to partition the graph.

Virtual vertices. The second method to incorporate the cost of migrating points uses the concept of *virtual vertices* [34]. For each processor the grid is enlarged with one (virtual) vertex, with edges connecting it with all vertices corresponding to grid points allocated on that processor. The resulting graph is then partitioned with the spectral bisection algorithm. Cutting an ordinary edge corresponds to the situation that two dependent grid points belong to different subgrids, whereas cutting a virtual edge means that a grid point is to be migrated to another processor. Just as in the clustering approach, some difficulties arise. The first one is that the extended graph should not be bisected such that more than one virtual vertex lies in one partition. Sometimes this is only possible at the price of some load imbalance, and will generally occur if the adapted grid differs very much from the original grid. Another problem is that the standard spectral bisection method does not always give good bisections for these extended graphs, in the sense that the new partition is almost the same as the old one, so that the load-imbalance is not removed. This is a consequence of the heuristic to relax the constraint $\xi_i \in \{-1, 1\}$ and to use the continuous formulation of the original optimization problem. In this case, the solution of the continuous optimization problem is no longer a good approximation to the solution of the discrete problem. Instead of solving the eigenproblem $L\xi = \lambda \xi$, it appears that it is better to scale L and solve the problem $L\xi = \lambda D\xi$ for the second smallest eigenvalue.

3.7. TRANSFER OF GRID POINTS

An essential part of regaining load-balance is to determine which grid points
are to be migrated between which processors. In the previous subsections
we have encountered this problem in two situations. In the first case, it
was clear which processors exchange how many grid points. This situation
occurred by the application of the diffusion method and the dimension
exchange method. In the multilevel method and the modifications of the
spectral bisection method, a number of grid points has to be transferred
between two groups of processors. In that case, it still must be determined
which processors exchange the grid points, and how many. We will briefly
consider these problems, following the approach of [36].

3.7.1. *Transfer between processor groups*

We assume that it is clear how many grid points should be transferred
from one group of processors to the other. Given these two groups $\mathcal{P}_1$
and $\mathcal{P}_2$, we will define the set of *border-processors* $\partial\mathcal{P}_1 \subset \mathcal{P}_1$ and $\partial\mathcal{P}_2 \subset$
$\mathcal{P}_2$ consisting of those processors which are connected by an edge of the
processor interconnection graph $\mathcal{T}$. We assume that the sets of border-
processors are not empty. If the total number of grid points to be transferred
from $\mathcal{P}_1$ to $\mathcal{P}_2$ is M, and the number of grid points allocated on processor
p is denoted by X_p, then each processor in $\partial\mathcal{P}_1$ will send

$$M_p = \frac{X_p}{\sum_{p \in \partial\mathcal{P}_1} X_p} M$$

grid points to some of the processors in $\partial\mathcal{P}_2$. Of course we have to assume
that this number M_p does not exceed X_p. For each processor p in $\partial\mathcal{P}_1$,
and each processor q in $\partial\mathcal{P}_2$, let ℓ_{pq} denote the number of grid points in
subdomain p which are connected to some grid point in subdomain q. Then
the number of grid points to be transferred from p to q can be taken as

$$M_{pq} = \frac{\ell_{pq}}{\sum_{q \in \partial\mathcal{P}_2} \ell_{pq}} M_p.$$

3.7.2. *Transfer between one pair of processors*

Once it has been determined how many grid points have to be transferred
from processor p to q, it must be decided which grid points actually are
involved in this transfer. One possibility to do this is to use the level-sets
as introduced in the beginning of this section. First select all grid points
in the first level-set L_1 which are connected to some point in subdomain q.
This will constitute a set S_1. In order to construct the set S_2, select all grid
points in the second level-set L_2 which are connected to some point in the

selected set S_1, etc. This selection algorithm stops if the number M_{pq} has been reached. Some care must be taken if some point in e.g. L_1 is connected to both a point in subdomain q_1 and to some point in subdomain q_2. The final set $S_1 \cup S_2 \cup \cdots$ is connected. This algorithm is based on connectivity information, and is thus most suitable to rebalance partitions obtained with a static method which is also based on connectivity information, such as e.g. the spectral bisection method. This approach is illustrated in Figure 10, where the first step of the diffusion method applied to the example in Figure 7 is shown in detail.

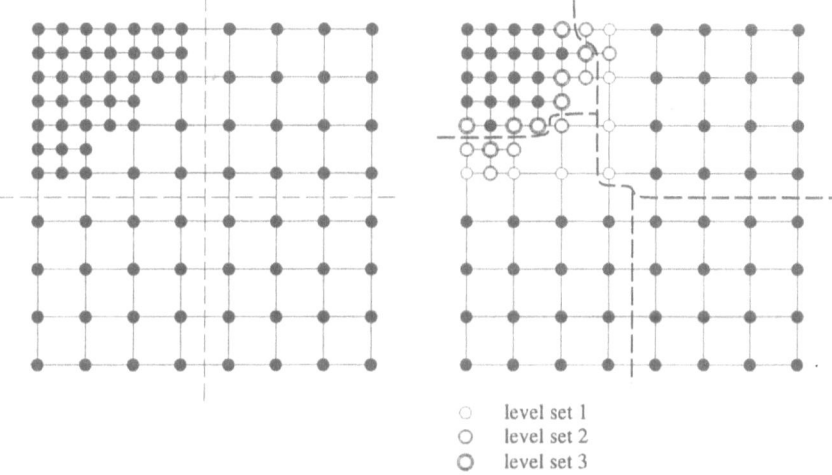

○   level set 1
○   level set 2
○   level set 3

Figure 10. Detailed execution of the first step of the diffusion method using connectivity information to designate the grid points to be transferred.

If the unbalanced partition originally comes from a partition obtained with one of the coordinate bisection methods, then a possibility to designate the grid points to be transferred is to use the coordinate variant of level-set. In this approach, grid points to be transferred are designated depending on their distance to the boundary between subdomains p and q. How to measure this distance depends on the particular bisection method. E.g. in the case of x-bisection, the x-coordinates of the grid points determine the distance. In Figure 11, we give an example where this approach gives better results than the designation based on connectivity-information.

References

1. B. Baldwin and H. Lomax. Thin layer approximation and algebraic model for separated turbulent flows, AIAA-78-257, 1978.

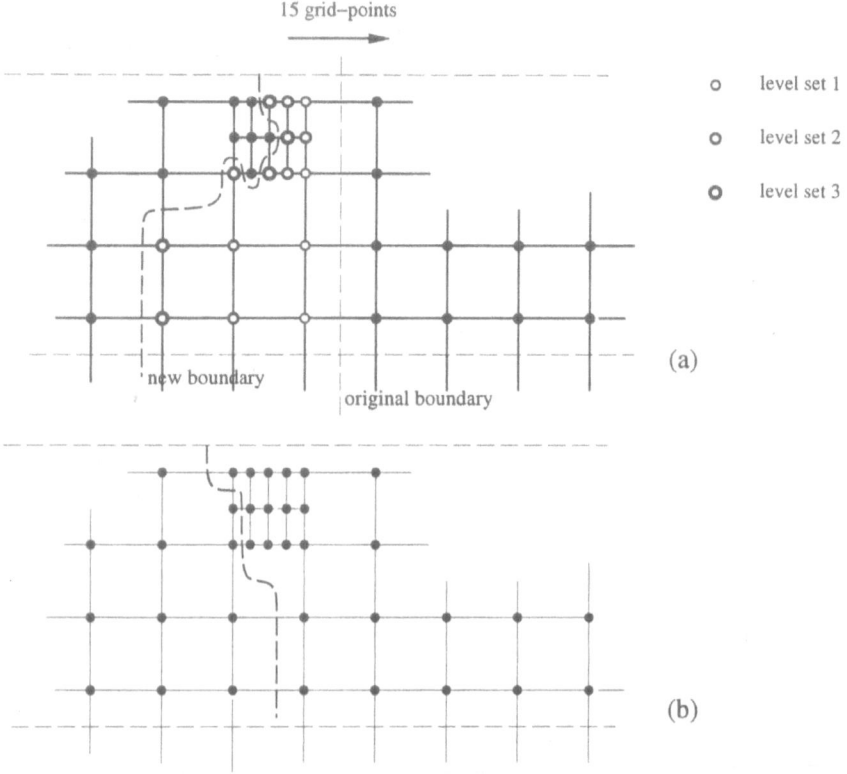

Figure 11.   An unbalanced grid is rebalanced by transferring 15 grid points from the left partition to the right. Shown are parts of these partitions. In Figure (a), connectivity-information is used to determine the points which are transferred. This leads to a jagged boundary. In Figure (b), the x-coordinate is used as a distance-estimator.

2. S.T. Barnard and H.D. Simon. Fast multilevel implementation of of recursive spectral bisection for partitioning unstructured problems. *Concurrency: practice and experience*, 6(2):101–117, 1994.

3. B.E. Bauer. *Practical parallel programming*. Academic press, San Diego, 1991.

4. J.E. Boillat. Load-balancing and poisson equation in a graph. *Concurrency: practice and experience*, 2(4):289–313, 1990.

5. J.E. Boillat, F. Brugè, and P.G. Kropf. A dynamic load-balancing algorithm for molecular dynamics simulation on multi-processor systems. *Journal of computational physics*, 96:1–14, 1991.

6. T. Bui, C. Heigham, C. Jones, and T. Leighton. Improving the performance of the Kernighan-Lin and simulated annealing graph bisection algorithm. In *26th ACM/IEEE design automation conference*, page 775. IEEE, 1989.

7. C.H. Cap and V. Strumpen. Efficient parallel computing in distributed workstation environments. *Parallel computing*, 19:1221–1234, 1993.

8. G. Cybenko. Dynamic load-balancing for distributed memory multiprocessors. *Journal of parallel and distributed computing*, 7:279–301, 1989.

9. H.L. deCougny et al. Load balancing for the parallel adaptive solution of partial

differential equations. *Applied numerical mathematics*, 16:157–182, 1994.

10. R. Van Driessche and D. Roose. Load balancing computational fluid dynamics calculations on unstructured grids. In *Special course on parallel computing in CFD*, number R-807 in AGARD report, pages 2-1 – 2-26, 1995.

11. Y.P. Chien et al. Dynamic load balancing on a network of workstations for solving computational fluid dynamics problems. *Computer methods in applied mechanics and engineering*, 119:17–33, 1994.

12. C. Farhat. A simple and efficient automatic FEM domain decomposer. *Computers and structures*, 28(5):579–602, 1988.

13. C. Farhat and M. Lesoinne. Automatic partitioning of unstructured meshes for the parallel solution of problems in computational mechanics. *International Journal for numerical methods in engineering*, 36:745–764, 1993.

14. C. Farhat and F.X. Roux. Implicit parallel processing in structural mechanics. *Computational mechanics advances*, 2:1–124, 1993.

15. R.D. Ferraro, P.C. Liewer, and V.K. Decyk. Dynamic load balancing for a 2d concurrent plasma PIC code. *Journal of computational physics*, 109:329–241, 1993.

16. M. Fiedler. Algebraic connectivity of graphs. *Czechoslovak mathematical journal*, 23(98):298–305, 1973.

17. M. Fiedler. A property of eigenvectors of nonnegative symmetric matrices and its application to graph theory. *Czechoslovak mathematical journal*, 25(100):619–633, 1975.

18. G.H. Golub and C.F. Van Loan. *Matrix computations*. Johns Hopkins University Press, Baltimore, MD, 1989.

19. G. Horton. A multi-level diffusion method for dynamic load-balancing. *Parallel computing*, 19(2):209–218, 1993.

20. B. Kernighan and S. Lin. An effective heuristic procedure for partitioning graphs. *Bell systems technical journal*, 49:291–308, 1970.

21. J. De Keyser, K. Lust, and D. Roose. Runtime load-balancing support for a parallel multiblock Euler/Navier-Stokes code with adaptive refinement on distributed memory computers. *Parallel computing*, 20:1069–1088, 1994.

22. J. De Keyser and D. Roose. run-time load balancing techniques for a parallel unstructured multi-grid euler solver with adaptive grid refinement. *Parallel computing*, 21:179–198, 1995.

23. R. Leland and B. Hendrickson. An empirical study of static load balancing algorithms. In *Proceedings of the scalable high-performance computing conference*, pages 682–685, 1994.

24. O.C. Martin and S.W. Otto. Partitioning of unstructured meshes for load balancing. *Concurrency: practice and experience*, 7(4):303–314, 1995.

25. A. Pothen, H.D. Simon, and K.-P Liou. Partitioning sparse matrices with eigenvectors of graphs. *SIAM Journal of Matrix Analysis and Applications*, 11(3):430–452, 1990.

26. J. Schabernack. Load balancing algorithms in distributed systems survey and taxonomy. *Informationstechnik (IT)*, 34(5):280–295, 1992.

27. H.D. Simon. Partitioning of unstructured problems for parallel processing. *Computing systems in engineering*, 2(2/3):135–148, 1991.

28. J. Song. A partially asynchronous and iterative algorithm for distributed load-balancing. *Parallel computing*, 20:853–868, 1994.

29. M. Streng, H.H. ten Cate, B.J. Geurts, and J.G.M. Kuerten. Simulation of compressible viscous flow on workstation clusters. Memorandum no. 1265, University of Twente, Enschede, 1995.

30. R. v. Hanxleden and L.R. Scott. Load balancing on message passing architectures. *Journal of parallel and distributed computing*, 13:312–324, 1991.

31. A.J. van der Steen, editor. *Aspects of computational science*. National Computing Facilities Foundation, The Hague, The Netherlands, 1995.

32. R. van Driessche and D. Roose. An efficient spectral bisection algorithm for dynamic

load balancing. Technical Report TW215, Department of computer science, K.U. Leuven, Belgium, 1994.

33. R. van Driessche and D. Roose. A graph contraction algorithm for the fast calculation of the Fiedler vector of a graph. In D.H. Bailey et al., editor, *Proceedings of the seventh SIAM Conference on parallel processing for scientific computing*, pages 621–626. SIAM, 1995.

34. R. van Driessche and D. Roose. An improved spectral bisection algorithm and its application to dynamic load balancing. *Parallel computing*, 21:29–48, 1995.

35. P.J.M. van Laarhoven and E.H.L. Aarts. *Simulated annealing: theory and applications*. D. Reidel publishing company, Dordrecht, The Netherlands, 1987.

36. A. Vidwans, Y. Kallinderis, and V. Venkatakrishnan. Parallel dynamic load-balancing algorithm for three-dimensional adaptive unstructured grids. *AIAA Journal*, 32(3):497–505, 1994.

37. C. Walshaw and M. Berzins. Dynamic load-balancing for PDE solvers on adaptive unstructured meshes. *Concurrency: practice and experience*, 7(1):17–28, 1995.

38. R.D. Williams. Performance of dynamic load balancing algorithms for unstructured mesh calculations. *Concurrency: practice and experience*, 3(5):457–481, 1991.

39. C.Z. Xu and F.C.M. Lau. Analysis of the generalized dimension exchange method for dynamic load-balancing. *Journal of parallel and distributed computing*, 16(4):385–393, 1992.

40. C.Z. Xu and F.C.M. Lau. Iterative load-balancing on multicomputers. *Journal of the operational research society*, 45(7):786–796, 1994.

41. C.Z. Xu, F.C.M. Lau, B. Monien, and R. Lüling. Nearest-neighbor algorithms for load-balancing in parallel computers. *Concurrency: practice and experience*, 7(7):707–736, 1995.

42. S. Zhou. *LSF: load sharing and batch queueing software*. Platform Computing Corporation, http://www.platform.com.

PARALLEL LINEAR SYSTEMS SOLVERS: SPARSE ITERATIVE METHODS

H. A. VAN DER VORST

Mathematical Institute
University of Utrecht
PO Box 80.010
NL-3508 TA Utrecht, the Netherlands

1. Introduction

Iterative methods are quite popular for the approximate solution of large sparse linear systems. As we will see, the are very well-suited for parallel computing. From this point of view we will discuss a number of methods, representative for the class of so-called Krylov subspace methods.

The Conjugate Gradient (CG) method is such an iterative method for the solution of linear systems Ax=b with symmetric positive definite A. Most of the operations in this method are trivially parallelizable. However, the innerproducts may spoil the performance of distributed memory computers when the number of processors is large. We will discuss alternative formulations for CG as well as possibilities to reschedule the operations, in an attempt to reduce negative effects on the performance due to communication. The matrix vector products may also introduce some communication overhead, but for many relevant problems this involves only communication with a few nearby processors. So this may, but does not necessarily, further degrade the performance of the algorithm.

The discussed approaches can also be used for related methods, such as Bi-CG, and hybrid methods like CGS and Bi-CGSTAB. The GMRES method is somewhat different, since the number of innerproducts increases linearly with the iteration count. This requires a slightly different approach.

These iterative methods are often used in combination with preconditioning in order to obtain much faster convergence. Popular preconditioners based upon incomplete LU decomposition of A are often problematic in a parallel environment. In our presentation we will give a brief overview of

P. Wesseling (ed.), High Performance Computing in Fluid Dynamics, 173–200.
© *1996 Kluwer Academic Publishers.*

parallelizable preconditioners, and of techniques to extract parallelism from
the classical incomplete LU decompositions.

In many scientific problems we are faced with the problem to solve very
large linear systems. Direct solution methods are often much too expensive
or they require too much memory space, and in these cases iterative solution methods offer an alternative. Usually these methods are rather modest
in their CPU and computer memory requirements per iteration step. However, the big disadvantage is that they are not robust, in the sense that
no particular iterative method is able to solve any given linear system to
a prescribed precision. The choice and tuning (preconditioning) of iterative methods for classes of linear problems is still a difficult and largely
unsolved problem. Nevertheless, for many of the very large systems they
are the only tools available for solution. We will not address the problem
of which method to select for a given class of linear systems.

Our starting point here is the assumption that iterative methods will be
used in a parallel environment and then the question remains how well iterative methods can take advantage of modern computer architectures. From
Dongarra's Linpack benchmark [20] it may be concluded that the solution
of a dense linear system can (in principle) be computed with computational
speeds close to peak speeds on most computers. This is already the case
for systems of, say, order 50000 on parallel machines with as many as 1024
processors.

In sharp contrast with the dense case are computational speeds reported
in [22] for the preconditioned as well as the unpreconditioned conjugate
gradient method (ICCG and CG, respectively).

In [22] a test problem was taken, generated by discretizing a three-
dimensional elliptic partial differential equation by the standard 7-point
central difference scheme over a three-dimensional rectangular grid, with
100 unknowns in each direction ($m = 100$, $n = 1,000,000$). The observed
computational speeds for several machines (1 processor in each case) are
given in Table 1.

We will focus our attention to some specific popular iterative methods
that can be considered, from a computational complexity point of view,
as archetypes for a wider class of methods, the so-called Krylov subspace
methods. In particular, we will discuss parallel properties of the Conjugate
Gradient method, GMRES, and Bi-CGSTAB(ℓ).

2. The conjugate gradients method

The standard (unpreconditioned) Conjugate Gradient algorithm [28, 26]
for the solution of $Ax = b$ may be represented by the following scheme:

x_0= initial guess; $r_0 = b - Ax_0$;

TABLE 1. : Speed in Megaflops for 50 Iterations.

Machine	optimized ICCG Mflops	Scaled CG Mflops	Peak Performance Mflops
NEC SX-3/22 (2.9 ns)	607	1124	2750
CRAY Y-MP C90 (4.2 ns)	444	737	952
CRAY 2 (4.1 ns)	96.0	149	500
IBM 9000 Model 820	39.6	74.6	444
IBM 9121 (15 ns)	10.6	25.4	133
DEC Vax/9000 (16 ns)	9.48	17.1	125
IBM RS/6000-550 (24 ns)	18.3	21.1	81
CONVEX C3210	15.8	19.1	50
Alliant FX2800	2.18	2.98	40

$$p_{-1} = 0; \widehat{\beta}_{-1} = 0;$$
$$\rho_0 = (r_0, r_0);$$
for $i = 0, 1, 2, ...$
$$\quad p_i = r_i + \widehat{\beta}_{i-1} p_{i-1};$$
$$\quad q_i = A p_i;$$
$$\quad \widehat{\alpha}_i = \frac{\rho_i}{(p_i, q_i)}$$
$$\quad x_{i+1} = x_i + \widehat{\alpha}_i p_i;$$
$$\quad r_{i+1} = r_i - \widehat{\alpha}_i q_i;$$
$\quad$ if x_{i+1} accurate enough then quit;
$$\quad \rho_{i+1} = (r_{i+1}, r_{i+1});$$
$$\quad \widehat{\beta}_i = \frac{\rho_{i+1}}{\rho_i};$$
end;

CG is most often used in combination with a suitable splitting $A = K - R$, and then K^{-1} is called the preconditioner. We will assume that K is also positive definite.

The CG method can be carried out for any choice of the innerproduct. Normally one takes the standard innerproduct $(x, y) = \sum x_i y_i$, but now we make a different choice:

$$[x, y] \equiv (x, Ky).$$

It is easy to verify that $K^{-1}A$ is symmetric positive definite with respect to $[,]$:

$$\begin{aligned}
[K^{-1}Ax, y] &= (K^{-1}Ax, Ky) = (Ax, y) \\
&= (x, Ay) = [x, K^{-1}Ay].
\end{aligned} \tag{1}$$

Hence, we can reformulate the CG procedure for solving the preconditioned system $K^{-1}Ax = K^{-1}b$, using the new [,]-innerproduct.

Whereas the standard CG method minimizes $(x_i - x, A(x_i - x))$, over all x_i in the Krylov subspace $\{r_0, Ar_0, ..., A^{i-1}r_0\}$, we are now minimizing the preconditioned system in the new norm:

$$[x_i - x, K^{-1}A(x_i - x)] = (x_i - x, A(x_i - x)),$$

which leads to the remarkable (and known) result that for this preconditioned system we still minimize the error in A-norm, although over a Krylov subspace generated by $K^{-1}r_0$ and $K^{-1}A$.

In the following computational scheme for preconditioned CG, for the solution of $Ax = b$ with preconditioner K^{-1}, we have replaced the [,]-innerproduct again by the familiar standard innerproduct. E.g., note that with $\tilde{r}_{i+1} = K^{-1}Ax_{i+1} - K^{-1}b$ we have that

$$\rho_{i+1} = [\tilde{r}_{i+1}, \tilde{r}_{i+1}]$$

$$= [K^{-1}r_{i+1}, K^{-1}r_{i+1}] = [r_{i+1}, K^{-2}r_{i+1}]$$

$$= (r_{i+1}, K^{-1}r_{i+1}),$$

and $K^{-1}r_{i+1}$ is the residual corresponding to the preconditioned system $K^{-1}Ax = K^{-1}b$.

$x_0 =$ initial guess; $r_0 = b - Ax_0$;
$p_{-1} = 0; \hat{\beta}_{-1} = 0$;
Solve w_0 from $Kw_0 = r_0$;
$\rho_0 = (r_0, w_0)$;
for $i = 0, 1, 2, ...$
    $p_i = w_i + \hat{\beta}_{i-1}p_{i-1}$;
    $q_i = Ap_i$;
    $\hat{\alpha}_i = \frac{\rho_i}{(p_i, q_i)}$
    $x_{i+1} = x_i + \hat{\alpha}_i p_i$;
    $r_{i+1} = r_i - \hat{\alpha}_i q_i$;
    if x_{i+1} accurate enough then quit;
    Solve w_{i+1} from $Kw_{i+1} = r_{i+1}$;
    $\rho_{i+1} = (r_{i+1}, w_{i+1})$;
    $\hat{\beta}_i = \frac{\rho_{i+1}}{\rho_i}$;
end;

Note that this formulation, which is quite popular, has the advantage that the preconditioner needs not to be split, and it is also avoided to back-transform solutions and residuals, as is necessary when one applies CG to $L^{-1}AL^{-1^T}y = L^{-1}b$.

2.1. PARALLELISM AND DATA LOCALITY IN PRECONDITIONED CG

Most often, the conjugate gradients method is used in combination with some kind of preconditioning. Usually, K is constructed as an approximation of A, such that systems like $Ky = z$ are much more easy to solve as $Ax = b$. Unfortunately, a popular class of preconditioners, based upon incomplete factorization of A, do not lend themselves very much for parallel implementation. We will discuss some approaches to obtain more parallelism in the preconditioner in section 5. For the moment we will assume that the preconditioner is chosen such that the parallelism in solving $Ky = z$ is comparable with the parallelism in computing Ap, for given p. The scheme for preconditioned CG is given in Section 2. Note that in that scheme the updating of x and r can only start after the completion of the innerproduct required for α_i. Therefore, this innerproduct is a so-called synchronization point: all further computation has to wait for completion of this operation. One can try to avoid such synchronization points as much as possible, or formulate CG in such a way that synchronization points can be combined.

Since on a distributed memory machine communication is required to assemble the innerproduct, it would be nice if we could proceed with other useful computation while the communication takes place. However, as we see from the CG scheme, there is no possibility to overlap this communication time with useful computation. The same observation can be made for the updating of p, which can only take place after the completion of the innerproduct for β_i. Apart from the computation of Ap, and the computations with K, we need to load 7 vectors for 10 vector floating point operations. This means that for this part of the computation only 10/7 floating point operation can be carried out per memory reference, in average.

Several authors ([8, 32, 33]) have attempted to improve this ratio, and to reduce the number of synchronization points. In our formulation of CG there are two such synchronization points. Meurant [32] (see also [39]) has proposed a variant in which there is only one synchronization point, however at the cost of a possibly reduced numerical stability, and one additional innerproduct. In this scheme the ratio between computations and memory references is about 2.

Chronopoulos and Gear [8] propose to further improve the data locality and parallelism in CG by combining s successive steps. Their algorithm is based upon the following property of CG. The residual vectors $r_0, ..., r_i$ form an orthogonal basis (assuming exact arithmetic) for the Krylov subspace spanned by $r_0, Ar_0, ..., A^{i-1}r_0$. When arrived at r_j, the vectors $r_0, r_1, ..., r_j$, $Ar_j, ..., A^{i-j-1}r_j$ also form a basis for this subspace. Hence, we may com-

bine s successive steps by generating $r_j, Ar_j, ..., A^{s-1}r_j$ first, and then do
the orthogonalization and the updating of the current solution with this
blockwise extended subspace. This approach leads to a modest increase in
flops in comparison with s successive steps of the standard CG.

The main drawback in this approach seems to be the potential numerical
instability. Depending on the spectral properties of A, the set $r_j, ..., A^{s-1}r_j$
may tend to converge to a vector in the direction of a dominating eigenvector, or, in other words, may tend to dependence for increasing values of s.
The authors claim to have seen successful completion of this approach, with
no serious stability problems, for small values of s. Nevertheless, it seems
that s-step CG, because of these problems, has a poor reputation (see also
[40]). However, a similar approach, suggested by Chronopoulos and Kim [9]
for other processes such as GMRES, seems to be more promising. Several
authors have pursued this research direction, and we will come back to this
in section 3.2.

In [18] a variant of CG is proposed, in which there is possibility to overlap all communication time with useful computations. This variant is a
rescheduled version of the original CG scheme, and is therefore precisely as
stable. The key trick in this approach is to delay the updating of the solution vector by one iteration step. Another advantage is that no additional
operations are required.

2.2. PARALLEL PERFORMANCE OF CG

From a parallel point of view, CG mimics very well parallel performance
properties of a variety of iterative methods such as Bi-CG, QMR, CGS,
Bi-CGSTAB, and others.

In this section we study the performance of CG on parallel distributed
memory systems and we report on some supporting experiments on actual
existing machines. Guided by our experiments we will discuss the suitability of CG for Massively Parallel Processing systems.

All computational intensive elements in preconditioned CG (updates, innerproducts, and matrix vector operations) are trivially parallelizable for
shared memory machines [21], except possibly for the preconditioning step:
Solve w_{i+1} *from* $Kw_{i+1} = r_{i+1}$. For the latter operation parallelism depends very much on the choice for K. In this section we restrict ourselves
to block Jacobi preconditioning, where the blocks have been chosen so that
each processor can handle one block independently of the others. For other
preconditioners that allow some degree of parallelism see [21] and Section 5.
For a distributed memory machine at least some of the steps require communication between processors: the accumulation of innerproducts and the
computation of Ap_i (depending on the non-zero structure of A and the

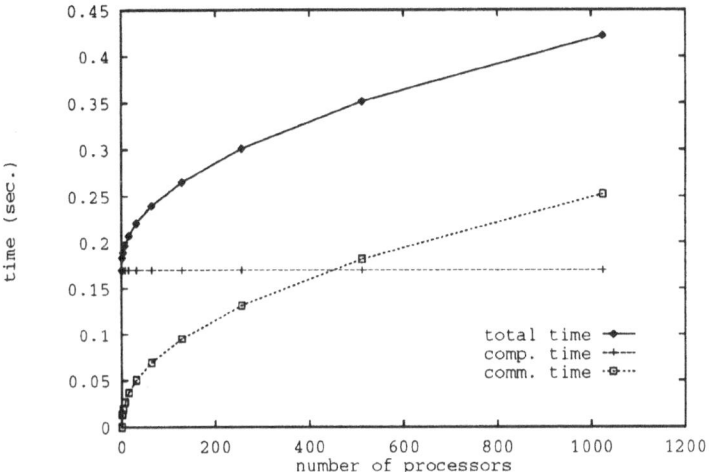

Fig.1: Modelled timings for 1 iteration with CG.

distribution of the non-zero elements over the processors). We consider in
some more detail the situation where A is a block-tridiagonal matrix of
order N, and we assume that all blocks are of order $\sqrt{N}$:

$$
A = \begin{pmatrix}
A_1 & D_1 & & & \\
D_1 & A_2 & D_2 & & \\
& D_2 & \ddots & \ddots & \\
& & & \ddots &
\end{pmatrix},
$$

in which the D_i are diagonal matrices, and the A_i are tridiagonal matrices. Such systems occur quite frequently in finite difference approximations
in 2 space dimensions. Our discussion can easily be adapted to 3 space
dimensions.

For simplicity we will assume that the processors are connected as a 2D
grid with $p \times p = P$ processors.
The data have been distributed in a straight forward manner over the processor memories and we have not attempted to fully exploit the underlying
grid structure for the given type of matrix in order to reduce communication as much as possible. In fact it will turn out that in our case the
communication for the matrix vector product plays only a minor role for
matrix systems of large size .
Because of symmetry only the 3 non-zero diagonals in the upper triangular

part of A need to be stored, and we have chosen to store successive parts
of length N/P of each diagonal in consecutive neighbouring processors.

The blocks for block Jacobi are chosen to be the diagonal blocks that
are available on each processor, and the various vectors (r_i, p_i, etc.) have
been distributed likewise, i.e. each processor holds a section of length N/P
of these vectors in its local memory. We will now inspect the ingredients of
CG for exploitable parallelism:

matrix vector product It is easily seen for a $2D$ processor grid (as well as
for many other configurations, including hypercube and pipeline), that the
matrix vector product can be completed with only neighbour-neighbour
communication. This means that the communication costs do not increase
for increasing values of p. If one follows a domain decomposition way of approach, in which the finite difference discretization grid is subdivided into p
by p subgrids (p in x-direction and p in y-direction), then the communication costs are smaller than the computational costs by a factor of $\mathcal{O}(\frac{\sqrt{N}}{p})$.
vector update In our case these operations do not require any communica-
tion and we should expect linear speed up when increasing the number of
processors P.

inner product For the innerproduct we need global communication for assembly and we need global communication for the distribution of the assembled innerproduct over the processors. For a $p \times p$ processor grid these
communication costs are proportional with p. This means that for a constant length of the vectorparts per processor, these communication costs
will dominate for values of p large enough. This is quite unlike the situation
for the matrix vector product and as we will see it may be a severely limiting
factor in achieving high speed-ups in a massively parallel environment.

On a 4-processor MEIKO SP1 we have done some experiments in order
to determine the costs of inter processor communication and for computation. Assuming that the costs for communication (for the innerproducts)
grow linearly with the length of the path of communication we have modelled the wall-clock time for a similar computer with P processors, for 1
iteration with CG with matrices of order $90000P$, as in Figure 1. Note that
we have increased the size of the linear system linearly with the number of
processors, which seems realistically since with larger computers one aims
to solve larger systems. The value 90000 has been chosen since this is more
or less the size of the part of the system that could be stored in the local
memory of one processor of our MEIKO machine.

From this Figure we learn that for P slightly larger than 400 the communication costs may be expected to dominate, and eventually they will
lead to very low speed-ups (even for systems for which the size is as large
as the total memory permits). We also see, that if overlap of communication and computation is possible, then potentially the communication can

be hidden for values of P less than 400, but this demands for a reformulation of the CG algorithm. Of course, these expectations are based on a model, but we have also carried out similar experiments on the 512 processor Parsytec GCel-3/512 of the University of Amsterdam [11]. In particular we have observed on that machine that the communication time for the innerproduct increases like $\sqrt{P}$, which just explains the behaviour of our model for the MEIKO-type of architecture.

If a given architecture permits the overlap of communication with computation, then we may try to reformulate CG in order to create possibilities for overlap. For the (extrapolated) MEIKO this may help for values of P up to about 400. For larger P we will see communication dominating anyhow, but the adverse effects can be lessened. A stable reformulation of CG which has this effect has been described in [18].

3. The minimum residual approach: GMRES and MINRES

The creation of an orthogonal basis for the Krylov subspace leads to

$$AV_i = V_{i+1}H_{i+1,i}. \tag{2}$$

We look for an $x_i \in K^i(A; r_0)$, that is $x_i = V_i y$, for which $\|b - Ax_i\|_2$ is minimal. This norm can be rewritten as

$$\|b - Ax_i\|_2 = \|b - AV_i y\|_2 = \|b - V_{i+1}H_{i+1,i}y\|_2.$$

Now we exploit the fact that V_{i+1} is an orthonormal transformation with respect to the Krylov subspace $K^{i+1}(A; r_0)$:

$$\|b - Ax_i\|_2 = \|\|r_0\|_2 e_1 - H_{i+1,i}y\|_2,$$

and this final norm can simply be minimized by solving the minimum norm least squares problem:

$$H_{i+1,i}y = \|r_0\|_2 e_1.$$

In GMRES [43] this is done with Givens rotations, that annihilate the subdiagonal elements in the upper Hessenberg matrix $H_{i+1,i}$.
Note that when A is symmetric the upper Hessenberg matrix $H_{i+1,i}$ reduces to a tridiagional system. This simplified structure can be exploited in order to avoid storage of all the basis vectors for the Krylov subspace, in a way similar as has been pointed out for CG. The resulting method is known as MINRES [35].

In order to avoid excessive storage requirements and computational costs for the orthogonalization, GMRES is usually restarted after each m iteration steps. This algorithm is referred to as GMRES(m); the not-restarted version is often called 'full' GMRES. Below we give a scheme for GMRES(m), that solves $Ax = b$, with a given preconditioner K.

x_0 is an initial guess;

for $j = 1, 2, \ldots$.

    Solve r from $Kr = b - Ax_0$;

    $v_1 = r/\|r\|_2$;

    $s := \|r\|_2 e_1$;

    for $i = 1, 2, \ldots, m$

        Solve w from $Kw = Av_i$;

        for $k = 1, \ldots, i$ orthogonalization of w

           $h_{k,i} = (w, v_k)$;   against v's, by modified

           $w = w - h_{k,i} v_k$;   Gram-Schmidt process

        end k;

        $h_{i+1,i} = \|w\|_2$;

        $v_{i+1} = w/h_{i+1,i}$;

        apply $J_1, \ldots, J_{i-1}$ on $(h_{1,i}, \ldots, h_{i+1,i})$;

    construct J_i, acting on i-th and $(i+1)$-st component

    of $h_{.,i}$, such that $(i+1)$-st component of $J_i h_{.,i}$ is 0;

        $s := J_i s$;

        if $s(i+1)$ is small enough then (UPDATE($\tilde{x}, i$); quit);

    end i;

    UPDATE($\tilde{x}, m$);

end j;

In this scheme UPDATE($\tilde{x}, i$) replaces the following computations:

Compute y as the solution of $Hy = \tilde{s}$

(in least squares sense if H is singular),

in which the upper i by i triangular part of H

has $h_{i,j}$ as its elements.

$\tilde{s}$ represents the first i components of s;

$\tilde{x} = x_0 + y_1 * v_1 + y_2 v_2 + \ldots + y_i v_i$;

s_{i+1} equals $\|b - A\tilde{x}\|_2$;

if this component is not small enough

then $x_0 = \tilde{x}$;

else quit;

There is an interesting and simple relation between the Ritz-Galerkin approach (FOM and CG) and the minimum residual approach (GMRES and MINRES). In GMRES the projected system matrix $H_{i+1,i}$ is transformed by Givens rotations to an upper triangular matrix (with last row equal to zero). So, in fact, the major difference between FOM and GMRES is that in FOM the last $((i+1)$-th row is simply discarded, while in GMRES this row is rotated to a zero vector. Let us characterize the Givens rotation, acting on rows i and $i+1$, in order to zero the element in position $(i+1, i)$, by the sine s_i and the cosine c_i. Let us further denote the residuals for FOM with an superscript F and those for GMRES with superscript G. Then the above observations lead to the following results for FOM and GMRES (for details see [43] and [7]).

1. The reduction for successive GMRES residuals is given by

$$\frac{\|r_k^G\|_2}{\|r_{k-1}^G\|_2} = |s_k| \qquad (3)$$

([43]: p. 862, Proposition 1).

2. If $c_k \neq 0$ then the FOM and the GMRES residuals are related by

$$\|r_k^G\|_2 = |c_k| \, \|r_k^F\|_2 \qquad (4)$$

([7]: theorem 5.1).

From these relations we see that when GMRES has locally a significant reduction in the norm of the residual (i.e., s_k is small), then FOM gives about the same result as GMRES (since $c_k^2 = 1 - s_k^2$). On the other hand when FOM has a breakdown ($c_k = 0$), then GMRES does not lead to an improvement in the same iteration step.

Because of these relations we can link the convergence behaviour of GMRES with the convergence of Ritz values (the eigenvalues of the "FOM" part of the upper Hessenberg matrix). This has been exploited in [53], for the analysis and explanation of local effects in the convergence behaviour of GMRES.

There are various different implementations of FOM and GMRES. Among those equivalent with GMRES are: Orthomin [56], Orthodir [29], Axelsson's method [1], and GENCR [25]. These methods are often more expensive than GMRES per iteration step. Orthomin seems to be still popular, since this variant can be easily truncated (Orthomin(s)), in contrast to GMRES. The truncated and restarted versions of these algorithms are not necessarily mathematically equivalent.

Methods that are mathematically equivalent with FOM are: Orthores [29] and GENCG [10, 58]. In these methods the approximate solutions are constructed such that they lead to orthogonal residuals (which form a basis for the Krylov subspace; analogously to the CG method). A good overview of all these methods and their relations is given in [42].

3.1. GMRESR AND GMRES⋆

In [54] it has been shown how the GMRES-method can be combined (or rather preconditioned) with other iterative schemes. The iteration steps of GMRES (or GCR) are called outer iteration steps, while the iteration steps of the preconditioning iterative method are referred to as inner iterations. The combined method is called GMRES⋆, where ⋆ stands for any given iterative scheme; in the case of GMRES as the inner iteration method, the combined scheme is called GMRESR[54].

Similar schemes have been proposed recently. In FGMRES[41] the update directions for the approximate solution are preconditioned, whereas in GM-RES⋆ the residuals are preconditioned. The latter approach offers more control over the reduction in the residual, in particular breakdown situations can be easily detected and remedied.

In exact arithmetic GMRES⋆ is very close to the Generalized Conjugate Gradient method[2]; GMRES⋆, however, leads to a more efficient computational scheme.

The GMRES⋆ algorithm can be described by the following computational scheme:

x_0 is an initial guess; $r_0 = b - Ax_0$;
for $i = 0, 1, 2, 3, \ldots$
    Let $z^{(m)}$ be the approximate solution of $Az = r_i$,
        obtained after m steps of an iterative method.
    $c = Az^{(m)}$ (often available from the iteration method)
    for $k = 0, \ldots, i-1$
        $\alpha = (c_k, c)$
        $c = c - \alpha c_k$
        $z^{(m)} = z^{(m)} - \alpha u_k$
    $c_i = c/\|c\|_2$; $u_i = z^{(m)}/\|c\|_2$
    $x_{i+1} = x_i + (c_i, r_i)u_i$
    $r_{i+1} = r_i - (c_i, r_i)c_i$
    if x_{i+1} is accurate enough then quit
end

A sufficient condition to avoid breakdown in this method ($\|c\|_2 = 0$) is that the norm of the residual at the end of an inner iteration is smaller than the norm of the right-hand side residual: $\|Az^{(m)} - r_i\|_2 < \|r_i\|_2$. This can easily be controlled during the inner iteration process. If stagnation occurs, i.e. no progress at all is made in the inner iteration, then it is suggested in [54] to do one (or more) steps of the LSQR method, which guarantees a reduction (but this reduction is often only small).

The idea behind this combined iteration scheme is that we explore parts of high-dimensional Krylov subspaces, hopefully localizing almost the same approximate solution that full GMRES would find over the entire subspace, but now at much lower computational costs. For the inner iteration we may select any appropriate solver, for instance, one cycle of GMRES(m), since then we have also locally an optimal method, or some other iteration scheme, like for instance Bi-CGSTAB.

In [16] it is proposed to keep the Krylov subspace, that is built in the inner iteration, orthogonal with respect to the Krylov basis vectors generated in the outer iteration. The procedure works as follows.

In the outer iteration process the vectors c_0, ..., c_{i-1} build an orthogonal basis for the Krylov subspace. Let C_i be the n by i matrix with columns c_0, ..., c_{i-1}. Then the inner iteration process at outer iteration i is carried out with the operator A_i instead of A, and A_i is defined as

$$A_i = (I - C_i C_i^T)A. \tag{5}$$

It is easily verified that $A_i z \perp c_0, ..., c_{i-1}$ for all z, so that the inner iteration process takes place in a subspace orthogonal to these vectors. The additional costs, per iteration of the inner iteration process, are i inner products and i vector updates. In order to save on these costs, one should realize that it is not necessary to orthogonalize with respect to all previous c-vectors, and that "less effective" directions may be dropped, or combined with others. In [16] suggestions are made for such strategies. Of course, these strategies are only attractive in cases where we see too little residual reducing effect in the inner iteration process in comparison with the outer iterations of GMRES⋆.

3.2. PARALLEL PERFORMANCE OF GMRES(M)

We will use a simple model for the computation time and the communication cost for the main kernels in GMRES(m); for a more elaborate description and analysis of the performance of Krylov subspace methods we refer to [15].

We use the term *communication cost* to indicate all the wall clock time spent in communication that is not overlapped with useful computation (so that it really contributes to wall-clock time).

The term *communication time* refers to the wall-clock time of the entire communication. In the case of a nonoverlapped communication, the communication time and the communication cost are the same.

Remark: Our quantitative formulas are not meant to give very accurate predictions of the exact execution times, but they will be used to identify the bottlenecks and to evaluate improvements.

Indeed, our experiments show that our models are relatively close to reality, and thus may be used as convenient tools for understanding the negative effects of global communication.

For a processor grid with $P = p^2$ processors, we have that $p_d = \sqrt{P}$. For nearest neighbour communication, let t_s denote the communication start-up time and let the transmission time associated with a single inner product computation be t_w. Then the global accumulation and broadcast time for

1 inner product is taken as $2p_d(t_s + t_w)$, while the global accumulation and broadcast for k simultaneous inner products takes $2p_d(t_s + k\, t_w)$.

For GMRES(m) the communication time for the modified Gram-Schmidt algorithm (with $\frac{1}{2}(m^2 + 3m)$ accumulations and broadcasts) is

$$T^G_{comm} = (m^2 + 3m)p_d(t_s + t_w). \qquad (6)$$

Note that for other processor configurations one only needs to bring in an appropriate value for p_d in (6).

From (6) we conclude that the communication cost for GMRES(m) is of the order $\mathcal{O}(m^2\sqrt{P})$ and for large processor grids this may become a bottleneck. Moreover, in the standard implementation we cannot reduce these costs by accumulating multiple inner products together, since the modified Gram-Schmidt orthogonalization of one single vector against a set of vectors and its subsequent normalization is an inherently sequential process.

However, if the modified Gram-Schmidt orthogonalization can be done for a set vectors simultaneously, then we have more possibilities for reduction of communication overhead. We discuss how to generate such a set of vectors in subsection 3.2.1.

Suppose the set of vectors $v_1, \hat{v}_2, \hat{v}_3 \ldots, \hat{v}_{m+1}$ has to be orthogonalized, where $\|v_1\|_2 = 1$. The modified Gram-Schmidt process can be implemented as is sketched in Figure 2. This reduces the number of messages to only m

> **for** $i = 1, 2, \ldots, m$ **do**
>   orthogonalize $\hat{v}_{i+1}, \ldots, \hat{v}_{m+1}$ on v_i;
> $v_{i+1} = \hat{v}_{i+1}/\|\hat{v}_{i+1}\|_2$;
> **end**

Fig.2: block-wise modified Gram–Schmidt.

instead of $\frac{1}{2}(m^2 + 3m)$ for the usual implementation of GMRES(m), but the length of the messages has increased. In this way, start-up time is saved by packing small messages, corresponding to one block of orthogonalizations, into one larger message. Note that the computation time for this approach is equal to that for the standard modified Gram–Schmidt algorithm. We will call this variant of GMRES: **parGMRES(m)**.

3.2.1. *Creation of the basis before orthogonalization*
In order to be able to use this parallel modified Gram–Schmidt algorithm in GMRES(m), a basis for the Krylov subspace has to be generated first.

$$\hat{v}_1 = v_1 = r/\|r\|_2$$
for $i = 1, 2, \ldots, m$ do
$$\hat{v}_{i+1} = \hat{v}_i - d_i A\hat{v}_i$$
end

Fig.3: Generation of a polynomial basis.

The idea to start with some non-orthogonal basis for the Krylov subspace, and then to orthogonalize this basis afterwards, was already suggested for the CG algorithm, referred to as s-step CG, in [8] for shared (hierarchical) memory parallel vector processors.

In [8] it is also reported that the s-step CG algorithm may converge slowly due to numerical instability for $s > 5$. In the parGMRES(m) algorithm stability seems to be much less of a problem since each vector is explicitly orthogonalized against all the other vectors, and we first generate a polynomial basis for the Krylov subspace, see [3, 14]. Furthermore, because of the restart, rounding errors made before restart can not have an accumulated effect on iterations carried out after the restart.

The basis vectors for the Krylov subspace $\hat{v}_i$ are generated as indicated in Figure 3, where the parameters d_i are chosen to keep the condition number of the matrix $[v_1, \hat{v}_2, \ldots, \hat{v}_{m+1}]$ sufficiently small. In order to achieve this, one might start in the first cycle with a suitable Chebychev recursion, for later cycles one may then take into account the GMRES information of the previous cycle. For more details on this, see [3], and [45]: Section 6.

3.2.2. *Experimental observations*

We present some experimental observations on the parallel performance of GMRES(30) and parGMRES(30) on a 400-processor Parsytec Supercluster. The T800-20 transputer processors of the Parsytec are connected in a fixed 20×20 mesh, of which arbitrary submeshes can be used. The transputer supports only nearest neighbor synchronous communication; more complicated communication has to be programmed explicitly. The communication rate is fast compared with the flop rate, but to current standards the T800 is a slow processor. For further details on our testing circumstances, see [17].

We will consider the performance of only one (par)GMRES(30) cycle.

We have solved a convection diffusion problem discretized by finite volumes over a 100×100 grid, resulting in the standard five-diagonal matrix with a tridiagonal block-structure, corresponding to the 5-point star. This relatively small problem size was chosen, because for processor grids of increasing size it very well shows the expected degradation of perfor-

TABLE 2. : Efficiencies and speed-ups.

| processor | GMRES(30) | | parGMRES(30) | |
grid	E (%)	S	E (%)	S
10×10	77.2	77.2	95.9	95.9
14×14	53.9	106.	88.8	174.
17×17	40.5	117.	69.4	201.
20×20	28.6	114.	53.4	214.

mance for GMRES(30) and the large improvements of parGMRES(30) over GMRES(30). Furthermore, the parallel behavior for this problem size on this slow machine corresponds quite well to much larger problems on more modern machines; see [17].

We give speed-ups and efficiencies in Table 2. Since the problem did not fit on a single processor, the speed-up and efficiency values were computed against a (fairly accurately) estimated sequential runtime.

We clearly see that the performance of parGMRES(30) is much better than the performance of GMRES(30). For a more elaborate discussion, including other values of m, see [17].

4. Bi-CGSTAB

Bi-CGSTAB [52] is based on the observation that the Bi-CG vector r_i is orthogonal to the entire subspace $K^i(A^*, w_1)$. As a result, we can, instead of squaring the Bi-CG polynomial, construct iteration methods, by which x_i are generated so that $r_i = \tilde{P}_i(A)P_i(A)r_0$ with other i^{th} degree polynomials $\tilde{P}$. An obvious possibility is to take for $\tilde{P}_j$ a polynomial of the form

$$Q_i(x) = (1 - \omega_1 x)(1 - \omega_2 x)...(1 - \omega_i x), \qquad (7)$$

and to select suitable constants ω_j. This expression leads to an almost trivial recurrence relation for the Q_i.

In Bi-CGSTAB ω_j in the jth iteration step is chosen as to minimize r_j, with respect to ω_j, for residuals that can be written as $r_j = Q_j(A)P_j(A)r_0$.

The preconditioned Bi-CGSTAB algorithm for solving the linear system $Ax = b$, with preconditioning K reads as follows:

x_0 is an initial guess; $r_0 = b - Ax_0$;
$\bar{r}_0 (= w_1)$ is an arbitrary vector, such that
$\quad (\bar{r}_0, r_0) \neq 0$, e.g., $\bar{r}_0 = r_0$;
$\rho_{-1} = \alpha_{-1} = \omega_{-1} = 1$;
$v_{-1} = p_{-1} = 0$;

for $i = 0, 1, 2, \ldots$
$\quad \rho_i = (\bar{r}_0, r_i); \beta_{i-1} = (\rho_i/\rho_{i-1})(\alpha_{i-1}/\omega_{i-1});$
$\quad p_i = r_i + \beta_{i-1}(p_{i-1} - \omega_{i-1} v_{i-1});$
$\quad$ Solve $\hat{p}$ from $K\hat{p} = p_i$;
$\quad v_i = A\hat{p};$
$\quad \alpha_i = \rho_i/(\bar{r}_0, v_i);$
$\quad s = r_i - \alpha_i v_i;$
$\quad$ if $\|s\|$ small enough then
$\quad\quad x_{i+1} = x_i + \alpha_i \hat{p};$ quit;
$\quad$ Solve z from $Kz = s$;
$\quad t = Az;$
$\quad \omega_i = (t, s)/(t, t);$
$\quad x_{i+1} = x_i + \alpha_i \hat{p} + \omega_i z;$
$\quad$ if x_{i+1} is accurate enough then quit;
$\quad r_{i+1} = s - \omega_i t;$
end

The matrix K in this scheme represents the preconditioning matrix and the way of preconditioning [52]. The above scheme carries out the Bi-CGSTAB procedure for the explicitly postconditioned linear system

$$AK^{-1}y = b,$$

but the vectors y_i and the residual have been backtransformed to the vectors x_i and r_i corresponding to the original system $Ax = b$. Compared with CGS, two extra innerproducts need to be calculated.

In exact arithmetic, the α_j and β_j have the same values as those generated by Bi-CG and CGS.

Bi-CGSTAB can be viewed as the product of Bi-CG and GMRES(1). Of course, other product methods can be formulated as well. Gutknecht [27] has proposed BiCGSTAB2, which is constructed as the product of Bi-CG and GMRES(2).

4.1. BI-CGSTAB2 AND VARIANTS

The residual $r_k = b - Ax_k$ in Bi-CG, when applied to $Ax = b$ with start x_0 can be written formally as $P_k(A)r_0$, where P_k is a k-degree polynomial. These residuals are constructed with one operation with A and one with A^* per iteration step. It was pointed out in [47] that with about the same amount of computational effort one can construct residuals of the form $\tilde{r}_k = P_k^2(A)r_0$, which is the basis for the CGS method. This can be achieved without any operation with A^*. The idea behind the improved efficiency of CGS is that if $P_k(A)$ is viewed as a reduction operator in Bi-CG, then one may hope that the square of this operator will be a twice as powerful reduction operator. Although this is not always observed in practice, one typically has that CGS converges faster than Bi-CG. This, together with the absence of operations with A^*, explains the success of the CGS method. A drawback of CGS is that its convergence behavior can look quite irregular, that is the norms of the residuals converge quite irregularly, and it may easily happen that $\|r_{k+1}\|_2$ is much larger than $\|r_k\|_2$ for certain k (for an explanation of this see [51]).

In [52] it was shown that by a similar approach as for CGS, one can construct methods for which r_k can be interpreted as $r_k = P_k(A)Q_k(A)r_0$, in which P_k is the polynomial associated with Bi-CG and Q_k can be selected free under the condition that $Q_k(0) = 1$. In [52] it was suggested to construct Q_k as the product of k linear factors $1 - \omega_j A$, where ω_j was taken to minimize locally a residual. This approach leads to the Bi-CGSTAB method. Because of the local minimization, Bi-CGSTAB displays a much smoother convergence behavior than CGS, and more surprisingly, it often also converges (slightly) faster. A weak point in Bi-CGSTAB is that we get breakdown if an ω_j is equal to zero. One may equally expect negative effects when ω_j is small. In fact, BiCGSTAB can be viewed as the combined effect of Bi-CG and GCR(1), or GMRES(1), steps. As soon as the GCR(1) part of the algorithm (nearly) stagnates, then the Bi-CG part in the next iteration step cannot (or only poorly) be constructed. For an analysis, as well as for suggestions to improve the situation, see [44].

Another dubious aspect of Bi-CGSTAB is that the factor Q_k has only real roots by construction. It is well-known that optimal reduction polynomials for matrices with complex eigenvalues may have complex roots as well. If, for instance, the matrix A is real skew-symmetric, then GCR(1) stagnates forever, whereas a method like GCR(2) (or GMRES(2)), in which we minimize over two combined successive search directions, may lead to convergence, and this is mainly due to the fact that then complex eigenvalue components in the error can be effectively reduced.

This point of view was taken in [27] for the construction of the BiCGSTAB2 method. In the odd-numbered iteration steps the Q-polynomial is expanded by a linear factor, as in Bi-CGSTAB, but in the even-numbered steps this linear factor is discarded, and the Q-polynomial from the previous even-numbered step is expanded by a quadratic $1 - \alpha_k A - \beta_k A^2$. For this construction the information from the odd-numbered step is required. It was anticipated that the introduction of quadratic factors in Q might help to improve convergence for systems with complex eigenvalues, and, indeed, some improvement was observed in practical situations (see also [36]).

However, our presentation suggests a possible weakness in the construction of BiCGSTAB2, namely in the odd-numbered steps the same problems may occur as in Bi-CGSTAB. Since the even-numbered steps rely on the results of the odd-numbered steps, this may equally lead to unnecessary break-downs or poor convergence. In [46] another, and even simpler approach was taken to arrive at the desired even-numbered steps, without the necessity of the construction of the intermediate Bi-CGSTAB-type step in the odd-numbered steps. Hence, in this approach the polynomial Q is constructed straight-away as a product of quadratic factors, without ever constructing a linear factor. As a result the new method Bi-CGSTAB(2) leads only to significant residuals in the even-numbered steps and the odd-numbered steps do not lead necessarily to useful approximations.

In fact, it is shown in [46] that the polynomial Q can also be constructed as the product of ℓ-degree factors, without the construction of the intermediate lower degree factors. The main idea is that ℓ successive Bi-CG steps are carried out, where for the sake of an A^*-free construction the already available part of Q is expanded by simple powers of A. This means that after the Bi-CG part of the algorithm vectors from the Krylov subspace $s, As, A^2s, ..., A^\ell s$, with $s = P_k(A)Q_{k-\ell}(A)r_0$ are available, and it is then relatively easy to minimize the residual over that particular Krylov subspace. There are variants of this approach in which more stable bases for the Krylov subspaces are generated [45], but for low values of ℓ a standard basis satisfies, together with a minimum norm solution obtained through solving the associated normal equations (which requires the solution of an ℓ by ℓ system. In most cases Bi-CGSTAB(2) will already give nice results for problems where Bi-CGSTAB or BiCGSTAB2 may fail. Note, however, that, in exact arithmetic, if no breakdown situation occurs, BiCGSTAB2 would produce exactly the same results as Bi-CGSTAB(2) at the even-numbered steps.

Bi-CGSTAB(2) can be represented by the following algorithm:

$$x_0 \text{ is an initial guess; } r_0 = b - Ax_0;$$
$$\hat{r}_0 \text{ is an arbitrary vector, such that } (r, \hat{r}_0) \neq 0,$$
$$\text{e.g., } \hat{r}_0 = r;$$
$$\rho_0 = 1; u = 0; \alpha = 0; \omega_2 = 1;$$
$$\text{for } i = 0, 2, 4, 6, ...$$
$$\rho_0 = -\omega_2 \rho_0$$

even BiCG step:
$$\rho_1 = (\hat{r}_0, r_i); \beta = \alpha\rho_1/\rho_0; \rho_0 = \rho_1$$
$$u = r_i - \beta u;$$
$$v = Au$$
$$\gamma = (v, \hat{r}_0); \alpha = \rho_0/\gamma;$$
$$r = r_i - \alpha v;$$
$$s = Ar$$
$$x = x_i + \alpha u;$$

odd BiCG step:
$$\rho_1 = (\hat{r}_0, s); \beta = \alpha\rho_1/\rho_0; \rho_0 = \rho_1$$

$$v = s - \beta v;$$
$$w = Av$$
$$\gamma = (w, \hat{r}_0); \alpha = \rho_0/\gamma;$$
$$u = r - \beta u$$
$$r = r - \alpha v$$
$$s = s - \alpha w$$
$$t = As$$

GCR(2)-part:
$$\omega_1 = (r, s); \mu = (s, s); \nu = (s, t); \tau = (t, t);$$
$$\omega_2 = (r, t); \tau = \tau - \nu^2/\mu; \omega_2 = (\omega_2 - \nu\omega_1/\mu)/\tau;$$
$$\omega_1 = (\omega_1 - \nu\omega_2)/\mu$$
$$x_{i+2} = x + \omega_1 r + \omega_2 s + \alpha u$$
$$r_{i+2} = r - \omega_1 s - \omega_2 t$$
if x_{i+2} accurate enough then quit
$$u = u - \omega_1 v - \omega_2 w$$

end

For more general Bi-CGSTAB(ℓ) schemes see [46, 45].

Another advantage of Bi-CGSTAB(2) over BiCGSTAB2 is in its efficiency. The Bi-CGSTAB(2) algorithm requires 14 vector updates, 9 innerproducts and 4 matrix vector products per full cycle. This has to be compared with a combined odd-numbered and even-numbered step in BiCGSTAB2, which requires 22 vector updates, 11 innerproducts, and 4 matrix vector products, and with two steps of Bi-CGSTAB which require 4 matrix vector products, 8 innerproducts and 12 vector updates. The numbers for BiCGSTAB2 are based on an implementation described in [36].

Also with respect to memory requirements, Bi-CGSTAB(2) takes an intermediate position: it requires 2 n-vectors more than Bi-CGSTAB and 2 n-vectors less than BiCGSTAB2.

For distributed memory machines the innerproducts may cause communication overhead problems (see, e.g., [12]). We note that the Bi-CG steps are very similar to conjugate gradient iteration steps, so that we may consider all kind of tricks that have been suggested to reduce the number of synchronization points caused by the 4 innerproducts in the Bi-CG parts. For an overview of these approaches see [4]. If on a specific computer it is possible to overlap communication with communication, then the Bi-CG parts can be rescheduled as to create overlap possibillities: 1. the computation of ρ_1 in the even Bi-CG step may be done just before the update of u at the end of the GCR part.

2. The update of x_{i+2} may be delayed until after the computation of γ in the even Bi-CG step.

3. The computation of ρ_1 for the odd Bi-CG step can be done just before the update for x at the end of the even Bi-CG step.

4. The computation of γ in the odd Bi-CG step has already overlap possi-

billities with the update for u.

For the GCR(2) part we note that the 5 innerproducts can be taken together, in order to reduce start-up times for their global assembling. This gives the method Bi-CGSTAB(2) a (slight) advantage over Bi-CGSTAB. Furthermore, we note that the updates in the GCR(2) may lead to more efficient code than for BiCGSTAB, since some of them can be combined.

5. Parallelism in the preconditioner

In this section we consider a number of possibilities to obtain parallelism in the standard Incomplete Choleski preconditioner [31]. The linear systems are supposed to arise from standard finite difference discretisations of second order pde's over rectangular grids in two or three dimensional space.

All of the discussed iterative methods may be combined with preconditioning. The easiest way to explain this is the following one. Given the linear system $Ax = b$, we construct an operator K that approximates A but that leads to simpler to solve systems. Often K is given in factored form (incomplete decompositions [31]): $K = LU$, and then we apply the iterative scheme for the preconditioned system $K^{-1}Ax = K^{-1}b$. Note that we can avoid the explicit inversion of K, since the operator $K^{-1}A$ is only used in matrix vector operations like

$$g = K^{-1}Ap.$$

The vector q is constructed in two steps:
1. $\tilde{p} = Ap$
2. Solve q from $Kq = \tilde{p}$.

Given the LU factorization of K, the second step can be carried out again in two successive operations:
2a. Solve z from $Lz = \tilde{p}$
2b. Solve q from $Uq = \tilde{z}$.

The lower triangular systems with L and U lead to back substitutions and this leads to major problems with parallelism (it also leads to a lesser extent to problems on vector computers).

In this section we will briefly overview some possibilities for exploiting or creating parallelism in the preconditioner. In particular, we will discuss parallelization techniques, including re-ordering, series expansion and domain decomposition techniques. Generally, the class of incomplete LU preconditioners does not possess a high degree of parallelism in its original form. Re-ordering and approximations by truncating certain series expansion will increase the parallelism, but usually with a deterioration in convergence rate. Domain decomposition offers a compromise.

There are two general approaches for parallelizing numerical methods:

1. extract maximal parallelism from a method which works well on sequential computer, *without* changing its numerical properties,
2. modify or approximate a good sequential method to increase the parallelism available, thus possibly degrading its numerical properties.

There is a fundamental difficulty when applying the above general principles to incomplete factorization methods. The major parallel bottleneck lies in the backsolves involving the triangular LU factors. The same bottleneck arises in computing the LU factors themselves but this occurs only once in the beginning of the iteration process. Even though these backsolves possess some degree of parallelism which can be exploited, this is often not sufficient to efficiently exploit many parallel architectures, especially massively parallel ones. On the other hand, modifying or approximating the sequential method in order to increase the amount of parallelism invariably leads to slower convergence rates. This should not be too surprising; it is just an instance of the fundamental trade-off between parallelism (which prefers locality) and fast convergence rate (which prefers global dependence) which governs many genuinely globally coupled systems (e.g. elliptic PDEs). The goal is to make the right trade-off for a given architecture - algorithm configuration.

There are three basic methodologies to extract or increase the parallelism in ILU methods: re-ordering, series expansions (including polynomial preconditioners), and domain decomposition. We shall briefly discuss them next.

5.1. RE-ORDERING

The Wavefront Ordering: Assume that the ILU factors have been computed and consider now the task of computing the product $y = L^{-1}v$. The goal is to find an ordering with which the components of y can be computed with maximal parallel efficiency. This can be accomplished by exploiting the dependency graph for the computations. A similar ordering can be used for computing $U^{-1}v$.

It is obvious that if one uses this idea for finite difference operators on a regular d dimensional grid with n grid points in each direction, then we can extract $O(n^{d-1})$ degree of parallelism. Thus, for a fixed number of processors, higher dimensional problems are easier to parallelise. On the other hand, potential degradation in performance can be caused by memory addressing with unequal stride for $d > 2$ and cache problems with the indirect addressing needed to access data on the wavefronts when the grid is stored as a 2D array.

The use of wavefront ordering has been investigated in [38] for the IBM 3090; numerical experiments on the CM2 can be found in [6]. The wavefront (or hyperplane) ordering for regular 3D grids has been described in detail for vector computers in [50]. Recently, in [5] a data mapping is described for the wavefront ordering in 3D useful for distributed memory architectures together with performance results for an 8-processor Cray Y-MP, a 128 processor Intel iPSC/860, and a 32K processor CM-2.

Multi-color Orderings: Since the degree of parallelism for ILU methods are limited in the natural ordering, a popular alternative is to use orderings that are designed to be more parallel. However, it must be emphasized that most of these orderings are *not* equivalent to the natural ordering, in the sense that the ILU factors computed using these are generally different from those generated using the natural ordering. Thus, the goal is to trade-off the relatively fast and well-understood convergence rate of the natural ordering for orderings with a high degree of parallelism.

An example is the well-known red-black ordering for 5-point stencils in 2D. Because the red points depend only on the black points but not on each other, they can all be updated in parallel. Thus the degree of parallelism is $n^2/2$, a substantial increase from $O(n)$ for the natural ordering. However, since the data dependence are completely local and there is no global sharing of information, the convergence rate is poor [24]. In fact, it can be shown that the condition number of the preconditioned system in the red-black ordering is only about 1/4 that of the *unpreconditioned* system for ILU, MILU and SSOR, with no asymptotic improvement as h tends to zero [30].

One way to strike a better balance between parallelism and fast convergence is to use more colors. In principle, since the different colors are updated sequentially, using more colors decreases the parallelism but increases the global dependence and hence the convergence. The key is to choose the number of colors to match the architecture. For example, In [19] up to 75 colors are used for a 76^2 grid on the NEC SX-3/14 resulting in a 2 Gflops performance, which is much better than that for the wavefront ordering.

DeLong and Ortega [13] exploit the fact that SOR works well in combination with red-black ordering, and they suggest to take a fixed number of SOR iterations as a preconditioner for GMRES or Bi-CGSTAB. This results in a highly parallel method that is quite competitive with ILU preconditioning in terms of matrix vector operations.

Multi-wavefront Orderings: A different approach to increase parallelism is to use several hyperplane wavefronts to sweep through the grid, the idea being that all wavefronts can be updated in parallel. For example, in [50] it was suggested to starte wavefronts from each of the four corners in a 2D rectangular grid, or from the eight corners in a 3D grid. Earlier, a similar idea was proposed in which the grid is divided into equal parts (e.g. halves or quadrants) and each part is ordered in its own natural ordering [34].

5.2. SERIES EXPANSIONS

Instead of using an ordering with more parallelism, a quite different approach to increase the parallelism in the naturally ordered ILU method is to replace it by an *approximation* which can be evaluated more efficiently in parallel.

In order to illustrate this, consider the computation of $(I - L)^{-1}v$, which is needed in applying the preconditioner. Here we have assumed without loss of generality that the diagonal entries of the lower triangular factor has been scaled to unity. It can be easily proved that if the spectral radius $\rho(L)$ satisfies $\rho(L) < 1$ and L is n by n strictly lower triangular, then we have the following finite expansion:

Neumann Expansion: $(I - L)^{-1}v = (I + L + L^2 + ... + L^{n-1})v.$

Note that $L^n = 0$. Each of the terms on the right-hand-side in the above expansion can be evaluated in parallel efficiently because they only involve repeated multiplication of v by sparse matrices. The idea is to then truncate the expansion but keeping enough terms so that the convergence rate is not too adversely affected.

In [49] this idea was applied for a truncated Neumann expansion to the diagonal blocks (which correspond to grid lines) in the point ILU factorization in order to increase the degree of vectorization.

Finally, a related method is the class of *polynomial preconditioners* in which A^{-1} is approximated by a low degree polynomial in A, chosen in some optimal manner [23]. In [55] it is shown how GMRES can often be effectively preconditioned by a Chebyshev matrix polynomial, for which the coefficients are obtained from eigenvalue approximations from a limited number of GMRES steps. In particular, the harmonic Ritz values have been employed as useful approximations and a surprisingly simple algorithm is presented in [55] for the computation of the Chebyshev parameters of the Chebyshev polynomial over a piecewise linear contour that encloses the eigenvalue approximations. Using this type of relatively expensive polynomial preconditioners often leads to a significant reduction in GMRES steps and hence the required communication-intensive innerproducts have a smaller degrading effect on the parallel performance of the preconditioned GMRES algorithm on distributed memory machines.

5.3. DOMAIN DECOMPOSITION

In this general approach, the physical domain or grid is decomposed into a number of overlapping or non-overlapping subdomains on each of which an independent incomplete factorization can be computed and applied in parallel. The main idea is to obtain more parallelism at the subdomain level rather than at the grid point level. Usually, the interfaces or overlapping region between the subdomains must be treated in a special manner. The advantage of this approach is that it is quite general and can be used with different methods used within different subdomains.

Radicati and Robert [37] used an algebraic version of this approach by computing ILU factors within overlapping block diagonals of a given matrix A. When applying the preconditioner to a vector v, the values on the overlapped region is averaged from the two values computed from the two overlapping ILU factors.

The approach of Radicati and Robert has been further perfectioned in [15], who studies the effects of overlap from the point of view of geometric domain decompositioning. He introduces artificial mixed boundary conditions on the internal boundaries of the subdomains. In [15]:Table 5.8 experimental results are shown for a decomposition in 20×20 slightly overlapping subdomains of a 200×400 mesh for a discretized convection-diffusion equation (5-point stencil). When taking ILU preconditioning for each subdomain, it is shown that the complete linear system can be solved by GMRES on a 400-processor distributed memory Parsytec system with an efficiency in the order of 80% (compared with $\frac{1}{400}$-th of the CPU time of ILU preconditioned GMRES for the unpartitioned system on 1 single processor).

In [48] interface conditions along subdomains are studied and continuity for the solution at the interface is forced up to some degree. It is proposed to include also mixed derivatives in these relations. The involved parameters can be determined locally by means of normal mode analysis, and they are adapted to the discretized problem. It is shown that the resulting domain decomposition method defines a standard iterative method for some splitting $A = M - N$, and the local coupling aimes at minimizing the largest eigenvalues of $I - AM^{-1}$. Of course, this method can be accelerated and impressive results for GMRES acceleration are shown in [48]. Some attention is paid to the case where the solutions for the subdomains are obtained in only modest accuracy per iteration step.

Recently, Washio and Hayami [57] employed a domain decomposition approach for a rectangular grid by which one step of SSOR is done for the interior part of each subdomain. In order to make this domain-decoupled SSOR more resemble the global SSOR, the SSOR iteration matrix for each subdomain is modified by premultiplying it with a matrix $(I - X_L)^{-1}$ and postmultiplying it by $(I - X_U)^{-1}$. The matrices X_L and X_U depend on the couplings between adjacent subdomains. In order to further improve the parallel performance, the inverses are approximated by low-order truncated Neumann series. Experimental results have been shown for a 32-processor NEC-Cenju distributed memory computer.

References

1. O. Axelsson. Conjugate gradient type methods for unsymmetric and inconsistent systems of equations. *Lin. Alg. and its Appl.*, 29:1–16, 1980.
2. O. Axelsson and P. S. Vassilevski. A black box generalized conjugate gradient solver with inner iterations and variable-step preconditioning. *SIAM J. Matrix Anal. Appl.*, 12(4):625–644, 1991.
3. Zhaojun Bai, Dan Hu, and Lothar Reichel. A Newton basis GMRES implementation. Technical Report 91-03, University of Kentucky, 1991.
4. R. Barrett, M. Berry, T. Chan, J. Demmel, J. Donato, J. Dongarra, V. Eijkhout, R. Pozo, C. Romine, and H. van der Vorst. *Templates for the Solution of Linear Systems: Building Blocks for Iterative Methods.* SIAM, Philadelphia, PA, 1994.
5. E. Barszcz, R. Fatoohi, V. Venkatakrishnan, and S. Weeratunga. Triangular systems for CFD applications on parallel architectures. Technical report, NAS Applied Research Branch, NASA Ames Research Center, 1994.
6. H. Berryman, J. Saltz, W. Gropp, and R. Mirchandaney. Krylov methods preconditioned with incompletely factored matrices on the CM-2. Technical Report 89-54, NASA Langley Research Center, ICASE, Hampton, VA, 1989.
7. P. N. Brown. A theoretical comparison of the Arnoldi and GMRES algorithms. *SIAM J. Sci. Statist. Comput.*, 12:58–78, 1991.
8. A. T. Chronopoulos and C. W. Gear. s-Step iterative methods for symmetric linear systems. *J. on Comp. and Appl. Math.*, 25:153–168, 1989.
9. A. T. Chronopoulos and S. K. Kim. s-Step Orthomin and GMRES implemented on parallel computers. Technical Report 90/43R, UMSI, Minneapolis, 1990.
10. P. Concus and G. H. Golub. A generalized Conjugate Gradient method for non-symmetric systems of linear equations. Technical Report STAN-CS-76-535, Stanford University, Stanford, CA, 1976.
11. G. C. (Lianne) Crone. The conjugate gradient method on the parsytec GCel-3/512. *FGCS*, 11:161–166, 1995.
12. L. Crone and H. van der Vorst. Communication aspects of the conjugate gradient method on distributed-memory machines. *Supercomputer*, X(6):4–9, 1993.
13. M. A. DeLong and J. M. Ortega. SOR as a Preconditioner. *Appl. Num. Math.*, 18:431–440, 1995.
14. E. de Sturler. A parallel variant of GMRES(m). In R. Miller, editor, *Proc. of the fifth Int.Symp. on Numer. Methods in Eng.*, 1991.
15. E. De Sturler. *Iterative methods on distributed memory computers.* PhD thesis, Delft University of Technology, Delft, the Netherlands, 1994.
16. E. De Sturler and D. R. Fokkema. Nested Krylov methods and preserving the orthogonality. In N. Duane Melson, T.A. Manteuffel, and S.F. McCormick, editors, *Sixth Copper Mountain Conference on Multigrid Methods*, volume Part 1 of *NASA Conference Publication 3324*, pages 111–126. NASA, 1993.
17. E. De Sturler and H.A. van der Vorst. Reducing the effect of global communication in GMRES(m) and CG on parallel distributed memory computers. *J. Appl. Num. Math.*, 1995.
18. J. Demmel, M. Heath, and H. van der Vorst. Parallel numerical linear algebra. In *Acta Numerica 1993.* Cambridge University Press, Cambridge, 1993.
19. S. Doi and A. Hoshi. Large numbered multicolor MILU preconditioning on SX-3/14. *Int'l J. Computer Math.*, 44:143–152, 1992.
20. J. J. Dongarra. Performance of various computers using standard linear equations software in a fortran environment. Technical Report CS-89-85, University of Tennessee, Knoxville, 1990.
21. J. J. Dongarra, I. S. Duff, D. C. Sorensen, and H. A. van der Vorst. *Solving Linear Systems on Vector and Shared Memory Computers.* SIAM, Philadelphia, PA, 1991.
22. J. J. Dongarra and H. A. van der Vorst. Performance of various computers using standard sparse linear equations solving techniques. *Supercomputer*, 9(5):17–29,

1992.

23. P. F. Dubois, A. Greenbaum, and G. H. Rodrigue. Approximating the inverse of a matrix for use in iterative algorithms on vector processors. *Computing*, 22:257–268, 1979.

24. I. S. Duff and G. A. Meurant. The effect of ordering on preconditioned conjugate gradient. *BIT*, 29:635–657, 1989.

25. H. C. Elman. *Iterative methods for large sparse nonsymmetric systems of linear equations*. PhD thesis, Yale University, New Haven, CT, 1982.

26. G. H. Golub and C. F. van Loan *Matrix Computations*. The Johns Hopkins University Press, Baltimore, 1989.

27. M. H. Gutknecht. Variants of BICGSTAB for matrices with complex spectrum. *SIAM J. Sci. Comput.*, 14:1020–1033, 1993.

28. M. R. Hestenes and E. Stiefel. Methods of conjugate gradients for solving linear systems. *J. Res. Natl. Bur. Stand.*, 49:409–436, 1954.

29. K. C. Jea and D. M. Young. Generalized conjugate-gradient acceleration of nonsymmetrizable iterative methods. *Lin. Algebra Appl.*, 34:159–194, 1980.

30. J.C.C. Kuo and T.F. Chan. Two-color fourier analysis of iterative algorithms for elliptic problems with red/black ordering. *SIAM J. Sci. Stat. Comp.*, 11:767–793, 1990.

31. J. A. Meijerink and H. A. van der Vorst. An iterative solution method for linear systems of which the coefficient matrix is a symmetric M-matrix. *Math.Comp.*, 31:148–162, 1977.

32. G. Meurant. The block preconditioned conjugate gradient method on vector computers. *BIT*, 24:623–633, 1984.

33. G. Meurant. Numerical experiments for the preconditioned conjugate gradient method on the CRAY X-MP/2. Technical Report LBL-18023, University of California, Berkeley, CA, 1984.

34. G. Meurant. Domain decomposition methods for partial differential equations on parallel computers. *Int. J. Supercomputing Appls.*, 2:5–12, 1988.

35. C. C. Paige and M. A. Saunders. Solution of sparse indefinite systems of linear equations. *SIAM J. Numer. Anal.*, 12:617–629, 1975.

36. Claude Pommerell. *Solution of large unsymmetric systems of linear equations*. PhD thesis, Swiss Federal Institute of Technology, Zürich, 1992.

37. G. Radicati di Brozolo and Y. Robert. Parallel conjugate gradient-like algorithms for solving sparse non-symmetric systems on a vector multiprocessor. *Parallel Computing*, 11:223–239, 1989.

38. G. Radicati di Brozolo and M. Vitaletti. Sparse matrix-vector product and storage representations on the IBM 3090 with Vector Facility. Technical Report 513-4098, IBM-ECSEC, Rome, July 1986.

39. Y. Saad. Practical use of polynomial preconditionings for the conjugate gradient method. *SIAM J. Sci. Stat. Comput.*, 6:865–881, 1985.

40. Y. Saad. Krylov subspace methods on supercomputers. Technical report, RIACS, Moffett Field, CA, September 1988.

41. Y. Saad. A flexible inner-outer preconditioned GMRES algorithm. *SIAM J. Sci. Comput.*, 14:461–469, 1993.

42. Y. Saad and M. H. Schultz. Conjugate Gradient-like algorithms for solving nonsymmetric linear systems. *Math. of Comp.*, 44:417–424, 1985.

43. Y. Saad and M. H. Schultz. GMRES: a generalized minimal residual algorithm for solving nonsymmetric linear systems. *SIAM J. Sci. Statist. Comput.*, 7:856–869, 1986

44. G. L. G. Sleijpen and H.A. Van der Vorst. Maintaining convergence properties of BICGSTAB methods in finite precision arithmetic. *Numerical Algorithms*, 10:203–223, 1995.

45. G. L. G. Sleijpen, H.A. Van der Vorst, and D. R. Fokkema. Bi-CGSTAB(ℓ) and other hybrid Bi-CG methods. *Numerical Algorithms*, 7:75–109, 1994.

46. G. L. G. Sleijpen and D. R. Fokkema. BICGSTAB(ℓ) for linear equations involving unsymmetric matrices with complex spectrum. *ETNA*, 1:11–32, 1993.

47. P. Sonneveld. CGS: a fast Lanczos-type solver for nonsymmetric linear systems. *SIAM J. Sci. Statist. Comput.*, 10:36–52, 1989.

48. K.H. Tan. *Local coupling in domain decomposition*. PhD thesis, Utrecht University, Utrecht, the Netherlands, 1995.

49. H. A. van der Vorst. A vectorizable variant of some ICCG methods. *SIAM J. Sci. Stat. Comput.*, 3:86–92, 1982.

50. H. A. van der Vorst. High performance preconditioning. *SIAM J. Sci. Statist. Comput.*, 10:1174–1185, 1989.

51. H. A. van der Vorst. The convergence behaviour of preconditioned CG and CG-S in the presence of rounding errors. In O. Axelsson and L. Yu. Kolotilina, editors, *Preconditioned Conjugate Gradient Methods*, Berlin, 1990. Nijmegen 1989, Springer Verlag. Lecture Notes in Mathematics 1457.

52. H. A. van der Vorst. Bi-CGSTAB: A fast and smoothly converging variant of Bi-CG for the solution of non-symmetric linear systems. *SIAM J. Sci. Statist. Comput.*, 13:631–644, 1992.

53. H. A. van der Vorst and C. Vuik. The superlinear convergence behaviour of GMRES. *JCAM*, 48:327–341, 1993.

54. H. A. van der Vorst and C. Vuik. GMRESR: A family of nested GMRES methods. *Num. Lin. Alg. with Appl.*, 1:369–386, 1994.

55. M.B. van Gijzen. *Iterative solution methods for linear equations in finite element computations*. PhD thesis, Delft University of Technology, Delft, the Netherlands, 1994.

56. P. K. W. Vinsome. ORTOMIN: an iterative method for solving sparse sets of simultaneous linear equations. In *Proc.Fourth Symposium on Reservoir Simulation*, pages 149–159. Society of Petroleum Engineers of AIME, 1976.

57. T. Washio and K. Hayami. Parallel block preconditioning based on SSOR and MILU. *Numer. Lin. Alg. with Applic.*, 1:533–553, 1994.

58. O. Widlund. A Lanczos method for a class of nonsymmetric systems of linear equations. *SIAM J. Numer. Anal.*, 15:801–812, 1978.

HIGH PERFORMANCE COMPUTING OF TURBULENT FLOWS

FUE-SANG LIEN
The University of Manchester Institute of Science and Technology
UMIST, PO Box 88, Manchester, M60 1QD, UK

1. Introduction

The quest for unlimited geometric flexibility as a prerequisite to the integration of CFD into the design cycle for real engineering components has led, over the past few years, to the development of flexible three-dimensional multi-block and unstructured-grid schemes supported by sophisticated grid-generation techniques. Current capabilities are such that quantitatively credible representations of flows around and within complex geometries can be attained by numerical computations, provided effects arising from turbulent transport do not contribute materially to the flow properties that govern important operational characteristics of the associated engineering configuration. This usually means that the boundary layers developing on its surface are thin and attached, and losses arising from turbulence are low and confined to a minor proportion of the whole flow domain. In contrast, for configurations such as wing-fuselage junctions and multi-stage turbomachines operating close to their 'off-design' conditions, the representation of turbulence effects can be of crucial importance to the predictive realism.

There are three distinctly different computational routes by which turbulence and/or its effects can be resolved: Direct Numerical Simulation (DNS), Large Eddy Simulation (LES) and Turbulence Modelling. The main difficulties associated with the first, and to some extent with the second as well, are the extremely high computing costs and memory requirements involved. The costs may be estimated on the basis of Kolomogorov length, velocity and time scales:

$$\eta = (\nu^3/\epsilon)^{1/4}, \quad v = (\nu\epsilon)^{1/4}, \quad \tau = (\nu/\epsilon)^{1/2}. \tag{1}$$

which are the smallest scales encountered in a turbulent flow.

If, say, 4 cells are required in each direction to fully resolve the eddy with length $\sim \eta$, then the total number of cells N (uniformly spaced) required

201

P. Wesseling (ed.), High Performance Computing in Fluid Dynamics, 201–236.
© *1996 Kluwer Academic Publishers.*

for DNS to solve a turbulent channel flow, with δ being half of the channel width, is:

$$N = [4\frac{2\delta}{\eta}]^3. \tag{2}$$

A typical value for $\epsilon \sim \frac{u_\tau^3}{2\delta}$ $(u_\tau = \sqrt{\tau_w/\rho})$. With this value substituted into equation (2), there results: $N \sim (100Re_\tau)^{9/4}$, in which $Re_\tau = \frac{u_\tau\delta}{\nu}$. The time interval required for credible turbulence statistics to be evaluated is approximately $100\delta/U_m$ and $U_m \sim 16u_\tau$. Assuming that $Re_\tau = 1,000$ and the time step $\Delta t \sim 0.002\delta/U_m$, the total numbers of cells and time steps required are approximately 1.8×10^{11} and 5×10^4, respectively.

LES resolves the large-to-medium scales of size larger than the mesh interval (typically, 1% of spatial dimension of the flow), and represents the largely dissipative effects of the small eddies by way of a Subgrid-Scale Model. Although current computer capabilities allow LES to be applied to a variety of relatively simple flows, resources for flows around complex configurations, say a high-lift wing with pylon and nacelle, at $Re_\infty = O(10^6)$ would require tens of thousands of CPU hours on today's supercomputers and are thus not tenable on economic grounds

Against the background of current research in DNS and LES, the consensus view is that modelling most engineering turbulent flows will continue to be based, well into the 21st century, on the solution of the Reynolds-averaged NS (RANS), energy and scalar-transport equations. Within this framework, the essential task is to construct a closure in which the unknown Reynolds stresses (and, in general, also the fluxes of heat and species concentrations) are related to known or determinable time- or ensemble-averaged flow properties and turbulence parameters. Conventional eddy-viscosity models based on the linear Boussinesq stress-strain relations are known to be afflicted by numerous weaknesses, including an inability to capture anisotropy, insufficient sensitivity to secondary strains, seriously excessive generation of turbulence at impingement zones and a violation of fundamental realisability constraints at large strain rates. Notwithstanding the above defects, eddy-viscosity models remain popular, and their use in complex flows is widespread due, principally, to their formalistic simplicity, numerical robustness, and computational economy. Second-moment closure, on the other hand, accounts for several of the key features of turbulence which are misrepresented by linear eddy-viscosity models, but is considerably more complex and can suffer from low numerical stability due to the lack of dominance of second-order fragments in the set of terms representing diffusion. As a result, CPU requirements for such closure models can be high, especially in 3D flows.

A potential alternative to second-moment closure that retains advantageous elements of the linear eddy-viscosity framework is one which is based on constitutive relations linking the Reynolds-stresses to non-linear expansions of strains and vorticity components. These may be cast in a form of additive terms, each pre-multiplied by an apparent viscosity - hence the term 'non-linear eddy-viscosity models' [Pope (1975), Gatski & Speziale (1993)]. Examples include the models of Speziale (1987), Shih et al (1993), Durbin (1995), Craft et al (1995) and Lien et al (1995a). The main differences between alternative variants of non-linear eddy-viscosity models are summarised in the Table 1, in which $A_2 = a_{ij}a_{ij}$ ($a_{ij} = \overline{u_i'u_j'}/k - 2/3\delta_{ij}$), and $\overline{v'^2}$ (which should be regarded as the turbulence intensity normal to the streamline) are obtained by solving related transport equations. The particular model variant used later in the present paper is that proposed by Lien et al (1995a), which is simple, easy to implement and performs well in both transitional and fully turbulent flows. Details of the model will be exposed in Section 3.1.

TABLE 1. Summary of recent non-linear eddy-viscosity models

Author(s)	Model form	Order in the stress-strain relationship	Number of turbulence transport equations
Speziale (1987)	High-Re	quadratic	2, $k - \epsilon$
Shih et al (1987)	High-Re	quadratic	2, $k - \epsilon$
Durbin (1995)	Low-Re	quadratic	3, $k - \epsilon - \overline{v'^2}$
Craft et al (1995)	Low-Re	cubic	3, $k - \epsilon - A_2$
Lien et al (1995)	Low-Re	cubic	2, $k - \epsilon$

Although rapid progress in computer technology, in terms of both CPU speed and memory, is inclined to favour LES as the approach taken to the computation of unsteady flows around configurations of industrial relevance, LES as well as DNS and RANS schemes are likely to continue to co-exists, each used to calculate different aspects of turbulent flows and different classes of flows. The question may then be posed what kind of computer architectures should be considered for performing the above types of computations in the foreseeable future. The first option is a vector computer, which, with state-of-the-art compiler technology, requires very little user interference and, yet, is able to generate results at a speed approximately 5-15 times that of a scalar machine given a fixed clock rate. Vector computing continues to be important in practice, and it is appropriate, therefore, to explain here briefly the basic concept of vectorisation and

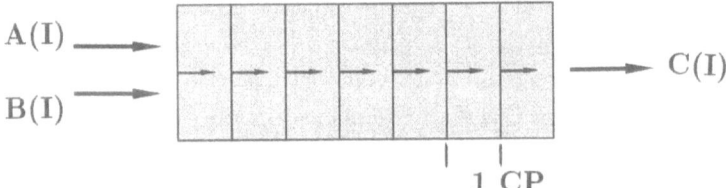

Figure 1.   Pipeling featured in vector hardware

highlight the supporting hardware by examining a simple DO-loop as executed, for example, on the CRAY Y-MP:

```
DO I=1,1000
C(I)=A(I)*B(I)
END DO
```

The operations involve three distinct steps:

1. Load arrays (or vectors) A and B from memory into vector register V0 and V1 (17 clock periods or "CP", henceforth).
2. Process the data in V0 and V1 by use of the Floating-point Multiply functional unit, storing result in register V2 (7CPs).
3. Store V2 in memory (17CPs).

The above load-multiply-store "assembly line" require 17+7+17=41 CPs in total to complete this operation, a sequence which is made possible by three important hardware features: *pipeling, chaining and vector registers.* The functional units are fully segmented, i.e. an operation is broken up into several one-clock-period segments. As a result, an operation like A(I)*B(I), which allows each step of the operation to pass its result to the next step after one CP, generates the first result after 7 CPs. Then, subsequent results can be obtained at the rate of one per CP. This process is called *pipeling*, and is shown schematically in Fig. 1. Because each function is designed for a special purpose, operations such as load, multiply and store can be performed independently. *Chaining* means that the output from one operation - say, multiply - can be used as a direct input for the next operation - say, store (see Fig. 2). Clearly, the efficiency of this vector processing will depend on the size of vector length, defined as the number of elements in a vector. The longer the vector the better. The maximum vector length on the CRAY Y-MP is 64 for each register, while other vector machines, such as the VP2000, can have vector length up to 1024.

In the presence of several nested DO loops, most vector computers will vectorise the innermost loop. For a long-vector-length machine, such as the

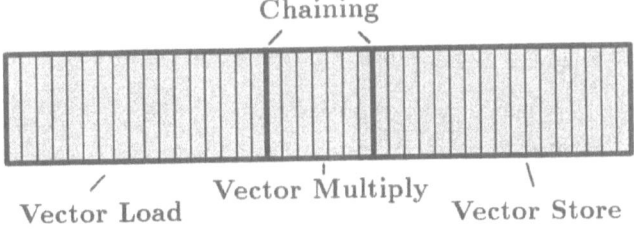

Figure 2. Chaining featured in vector hardware

VP2000, it is particularly advantageous to collapse nested loops into a single loop, i.e. a conversion of multi-dimensional arrays into one-dimensional arrays, in order to increase the vector length. On the other hand, for a machine like the CRAY Y-MP, loop collapse is not always necessary, especially in view of considerable programming effort involved. The CRAY complier *cf77* will automatically examine the loops from the innermost one outwards in order to identify an appropriate loop for vectorisation, unless user directives are inserted. However, the potential danger in doing so is the creation of a *memory contention* problem. The CRAY Y-MP uses banked shared memory, which can have up to 128 Mwords, arranged in 128 banks, say. [1]. Consider the following loops:

```
S------< DO J=1,128
S V----< DO I=1,400
S V        A(I)=A(I)+B(J,I)
S V----< END DO
S V----< END DO
```

Fig. 3 shows how consecutive elements of array B are stored in the memory bank, and demonstrates that memory conflicts occur by the vector load operation used to successively access the array element B(J,1:400) at the same memory bank. Since each access to the same location in a memory bank can delay the next access for a number of CPs, a vector load (or store) without memory contention runs, in general, 5 to 8 times faster than that with the most serious memory contention.

The speed-up ratio, s_v, resulting from vectorisation can be estimated

---

[1] Enter the "target" command to find out how many memory banks are configured on the system.

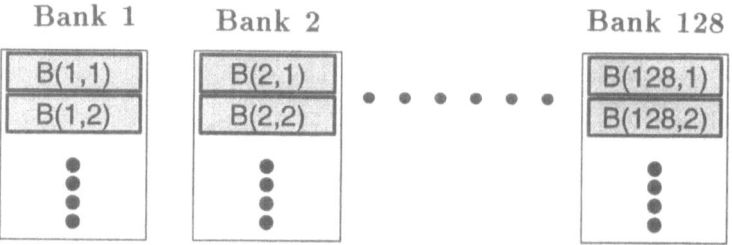

Figure 3.   Arrangement of memory bank on the CRAY Y-MP

by use of Amadahl's law:

$$s_v = \frac{1}{(1 - f_v) + \frac{f_v}{R_v}},$$ (3)

where

f_v = fraction of a programme that is vectorised

R_v = ratio of scalar to vector processing time

According to (3), $s_v \sim 5.3$ for a 90%-vectorised programme running on the CRAY, in which $R_v \sim 10$. One reason inhibiting a 100% vectorisation level is that real programs always contains algorithms, which are data-dependent and therefore intrinsically scalar. For example, a compilation listing[2] of a FORTRAN code relating to a Gauss-Seidel iteration scheme is given in Fig. 4. This is used to solve a 2D Poisson equation. Because the array element PHI(I,J) in the "DO I" loop requires PHI(I-1,J), which was defined in the previous iteration, the code involves what is termed *recurrence*. Recurrence would normally yield incorrect results when a reordering of operations occurs. As is recognised in Fig. 4(a), the "DO I" loop cannot be readily vectorised. One approach to overcoming this problem is to resort to the "Red & Black" method, which simply splits the "DO I" loop into two parts, with the stride set to 2. With this, the dependence of the array element at location (I,J) on that at (I-1,J) can be avoided, enabling the "DO I" loop to be vectorised, as is demonstrated in Fig. 4(b).

However ingenious one may be in exploiting vectorisation, a single CPU poses inevitable constraint of computing power and memory, and this motivates the migration to multi-processor systems, allowing tasks to be executed in parallel. Parallel computers can, in general, be classified into two

---

[2]The output is produced with the CRAY *cf77* complier with *-Wf"-em"* switch.

Gauss Seidel iteration for solving:

$$A_P \phi_P = \sum_{m=E,W,N,S} A_m \phi_m + S_U$$

(a)

```
S---------------<       DO J=2,NJM1
S                       IF (NIM1 - 1 .GE. 10) THEN
S                   CDIR@    IVDEP
S V----<               DO I = 1, NIM1 - 1
S  V                   R1V(I) = AE(1+I,J)*PHI(2+I,J  ) + AN(1+I,J)*PHI(1+I,J+1)
S  V          1               + AS(1+I,J)*PHI(1+I,J-1) + SU(1+I,J)
S  V---->               END DO
S S----<                DO I = 1, NIM1 - 1
S S  r                     PHI(1+I,J) = R1V(I) + AW(1+I,J)*PHI(I,J)
S S  r                     PHI(1+I,J) = PHI(1+I,J)/AP(1+I,J)
S S-r-->                END DO
S                       ELSE
S            CDIR@      NEXTSCALAR
S S----<                DO I = 2, NIM1
S S                        PHI(I,J) = AE(I,J)*PHI(I+1,J) + AW(I,J)*PHI(I-1,J)
S S          1                     + AN(I,J)*PHI(I,J+1) + AS(I,J)*PHI(I,J-1)
S S          1                     + SU(I,J)
S S                        PHI(I,J) = PHI(I,J)/AP(I,J)
S S---->                END DO
S                       ENDIF
S------>                END DO
```

(b)

```
                    C
                            L=2
S-----------------------< 100  CONTINUE
S S---------------------<       DO J=2,NJM1
S S                         CDIR@ IVDEP
S S V-------------------<       DO I=L,NIM1,2
S S V                             PHI(I,J)=
S S V                           & AE(I,J)*PHI(I+1,J)+AW(I,J)*PHI(I-1,J)
S S V                           &+AN(I,J)*PHI(I,J+1)+AS(I,J)*PHI(I,J-1)
S S V                           &+SU(I,J)
S S V                             PHI(I,J)=PHI(I,J)/AP(I,J)
S S V------------------->       END DO
S S--------------------->       END DO
S                           C
S                               IF(L.EQ.2) THEN
S                               L=L+1
S----------------------->       GO To 100
                                END IF
                            C
```

Figure 4. Compliation listing of Gauss-Seidel iteration: (a) without (b) with "Red and Black" method.

categories, depending on their memory architecture. The first category is characterised by a small number of very powerful vector processors sharing a common memory, which can either be split into different "banks" or connected to the processors with a high-speed "bus". The second category consists of a large number of less powerful processors with distributed memory, portions of which are physically attached to related processors. The processors cooperate and communicate by sending messages to and receiving messages from one another through a switching network. There are many possibilities in choosing the layout (or topology) of the communication network, leading to a wide diversity among the distributed-memory systems. Examples for (virtual) shared-memory systems are the CRAY Y-MP and C-90, CONVEX C2 and C3 and VPP500. Machines such as the Intel iPSC/860 hypercube, the IBM SP/2 and the CRAY T3D, belong to the category of distributed-memory systems.

A different classification of parallel computer system, widely used within the computing community, is one which distinguishes multiple-instruction-stream/multiple-data-stream (MIMD) from single-instruction-stream/multiple-data-stream (SIMD). Here, a "stream" refers to the sequence of data or instructions as seen by the machine during the execution of a programme. All of the systems mentioned above are belong to the MIMD category. On the other hand, SIMD machine, of which the Connection Machine is an example, always executes the same instruction at the same time, but on different data.

For shared-memory machines, parallelism is mainly directed, at least as seen by the user, towards DO-loop operations. Thus, a set of arrays is generally split into smaller chunks to which each processor applies an identical set of operations. However, the DO-loop level of parallelisation can, like that of vectorisation, be hindered by features linked to data dependency. Hence, recursive algorithms neither vectorise nor parallelise well on a shared-memory vector-parallel machine. One route to alleviating this difficulty will be considered in Section 4.1.

Alternatively, parallelisation can easily be achieved by use of the "domain decomposition" technique. In general, this involves three steps: (1) partitioning of the solution domain; (2) performing computations on each processor to update its own data set; (3) communicating data between processors. In order to implement this method within a structured-grid environment, the construction of connectivity matrices is necessary. This is used to provide topological information for each subdomain, allowing inter-subdomain communication to be effected correctly. This approach is hereafter referred to as the "multi-block" method.

The objective of this paper is to expose the details of an effort in which a multi-block serial code (STREAM, Lien et al, 1995b) has been ported to various parallel systems, including the CRAY Y-MP, T3D and the Intel iPSC/860. The basic numerical framework and, associated with it, a multi-block extension will be considered in Section 2. The turbulence model used in the computational studies to follow is described in Section 3, together with issues concerning with its numerical implementation. In Section 4, computational performance, especially parallel efficiency on both shared- and distributed-memory machines, will be examined, initially for a inviscid supersonic flow over a 2D circular bump, for which three types of communication routines have been adopted. Here, key elements of modifications introduced into the code are exemplified by means of a FORTRAN listing. Then, results are presented for a subsonic turbulent flow over a 3D ellipsoidal body at high incidence, performed with three turbulence models on a mesh containing up to 128^3 grids. Finally, Section 5 contains closing remarks and conclusions on the topics covered in this paper.

2. Description of a Multi-block Algorithm

2.1. BASIC NUMERICAL FRAMEWORK

The multi-block strategy takes as its starting point a single-block scheme which solves the Reynolds-averaged Navier-Stokes equations, the mass-conservation law and the appropriate turbulence-model equations. The finite-volume discretisation procedure and the implementation of turbulence models up to full second-moment closure within the single-block algorithm for both incompressible and compressible (transonic) flows have been presented in detail by Lien & Leschziner (1994a). Here, only those facets are summarised that impinge on the multi-block extension to follow in Section 2.2.

Any transport equation governing a flow property ϕ can be written, in terms of the curvilinear coordinate systems (ξ, η, ζ), as follows:

$$\frac{\partial}{\partial \xi}[U\phi - Jq_{11}^{\phi}\frac{\partial \phi}{\partial \xi}] + \frac{\partial}{\partial \eta}[V\phi - Jq_{22}^{\phi}\frac{\partial \phi}{\partial \eta}] + \frac{\partial}{\partial \zeta}[W\phi - Jq_{33}^{\phi}\frac{\partial \phi}{\partial \zeta}] = JS_{\phi}, \quad (4)$$

where the contravariant velocities (U, V, W) are given by:

$$U = J(u\xi_x + v\xi_y + w\xi_z), \quad V = J(u\eta_x + v\eta_y + w\eta_z), \quad W = J(u\zeta_x + v\zeta_y + w\zeta_z), \quad (5)$$

the coefficients q_{11}, q_{22}, q_{33} and Jacobian, J, are:

$$q_{11} = \xi_x\xi_x + \xi_y\xi_y + \xi_z\xi_z, \quad q_{22} = \eta_x\eta_x + \eta_y\eta_y + \eta_z\eta_z, \quad q_{33} = \zeta_x\zeta_x + \zeta_y\zeta_y + \zeta_z\zeta_z.$$

$$J = x_\xi y_\eta z_\zeta + x_\zeta y_\xi z_\eta + x_\eta y_\zeta z_\xi - x_\xi y_\zeta z_\eta - x_\zeta y_\eta z_\xi - x_\eta y_\xi z_\zeta. \quad (6)$$

Integration of Eq. (4) over the volume shown in Fig. 5, and application of the Gauss Divergence Theorem results in a balance of convective and diffusive cell-face fluxes and volume-integrated net source. The introduction of approximations which link the convective and diffusive fluxes to nodal values leads to a weighted-average formula of the form:

$$A_P\phi_P = \sum_{m=E,W,N,S,T,B} A_m\phi_m + S_\phi + (\frac{\rho^o J}{\Delta t})_P\phi_P^o \quad (7)$$

where ϕ stands either for momentum components or for any intensive scalar property, including transported turbulence quantities, and S_ϕ represents the source/sink term.

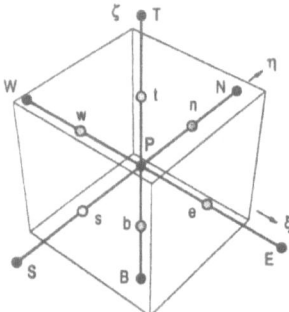

Figure 5. Finite volume and stroage arrangement

In the present implementation, either the QUICK scheme of Leonard (1979) or the UMIST-TVD scheme of Lien & Leschziner (1994b) have been used for convection, while the diffusive fluxes have been approximated by central differences. Whatever approximations are adopted for transport, Eq. (7) applies provided the source term is made to accommodates all links to nodes lying beyond the compact 7-point stencil P, E, W, N, S, T, B. Any higher-order convection scheme is here implemented as a first-order upwind approximation augmented by deferred corrections which are lumped in the source term S_ϕ. The coefficients A_m thus become:

$$A_E = (\Gamma_\phi J q_{11})_e + \text{MAX}[-(\rho U)_e, 0], \quad A_W = (\Gamma_\phi J q_{11})_w + \text{MAX}[(\rho U)_w, 0],$$

$$A_N = (\Gamma_\phi J q_{22})_n + \text{MAX}[-(\rho V)_n, 0], \quad A_S = (\Gamma_\phi J q_{22})_s + \text{MAX}[(\rho V)_s, 0],$$

$$A_T = (\Gamma_\phi J q_{33})_t + \text{MAX}[-(\rho W)_t, 0], \quad A_B = (\Gamma_\phi J q_{33})_b + \text{MAX}[(\rho W)_b, 0],$$

$$A_P = \sum_{m=E,W,N,S,T,B} A_m + (\frac{\rho J}{\Delta t})_P \tag{8}$$

With the UMIST-TVD used for convection, the deferred-corrector source in the ξ-direction, $S_\phi^{DC-\xi}$, is:

$$2S_\phi^{DC-\xi} = [(\rho U)_e^+ \varphi(r_e^+) - (\rho U)_e^- \varphi(r_e^-)](\phi_E - \phi_P)$$

$$- [(\rho U)_w^+ \varphi(r_w^+) - (\rho U)_w^- \varphi(r_w^-)](\phi_P - \phi_W), \tag{9}$$

in which $\varphi(r)$ is defined as:

$$\varphi(r) = \max[0, \min(2r, 0.25 + 0.75r, \underbrace{0.75 + 0.25r}_{\text{QUICK}}, 2)] \tag{10}$$

$$(\rho U)^\pm = \frac{\rho U \pm |\rho U|}{2}, \tag{11}$$

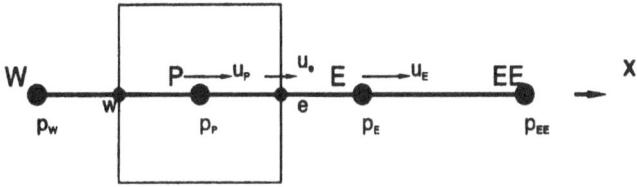

Figure 6. Grid stencil in ξ-direction for UMIST scheme

and $r_{e,w}^{\pm}$ (see Fig. 6) are

$$r_e^+ = \frac{\phi_P - \phi_W}{\phi_E - \phi_P}, \quad r_e^- = \frac{\phi_E - \phi_{EE}}{\phi_P - \phi_E}, \quad r_w^+ = \frac{\phi_W - \phi_{WW}}{\phi_P - \phi_W}, \quad r_w^- = \frac{\phi_P - \phi_E}{\phi_W - \phi_P},$$
(12)

Similar expressions of (9)-(12) apply to $S_\phi^{DC-\eta}$ and $S_\phi^{DC-\zeta}$.

To impose mass conservation, both in incompressible and compressible conditions, the SIMPLE pressure-correction algorithm of Patankar (1980) is adopted. Checkerboard oscillation arising from the collocated storage arrangement can be avoided by use of Rhie & Chow's interpolation (1983). Although this scheme is well known, its principal elements need to be highlighted here, at least in simple terms, to aid understanding of a particular facet of the multi-block extension to follow. The essential arguments may be conveyed by reference to a simple one-dimensional form of the discretised equation pertaining to the cell in Fig. 6. The key point is that the face velocity u_e is evaluated by linear interpolation of the momentum equations governing the nodal velocities u_P and u_E from which then the pressure gradients terms are subtracted and to which a compensating pressure-gradient fragment is finally added, the last formed only with the two nodal pressures straddling the face velocity u_e. The result is then:

$$u_e = \underbrace{\frac{1}{2}(u_P + u_E)}_{\text{linear interpolation}}$$

$$\underbrace{+ \frac{1}{2}\{[(\frac{DU}{A_P})_P + (\frac{DU}{A_P})_E](p_P - p_E) - [\frac{DU}{A_P}(p_w - p_e)]_P - [\frac{DU}{A_P}(p_w - p_e)]_E\}}_{\text{pressure smoothing}}$$
(13)

where DU_m is the u-directed cross-sectional area at location m ($m = P, E$). An important point to underline in relation to Eq. (13) is that the face velocity u_e depends upon pressure values at four nodes, two on either side of the face. This has implications as regards inter-block connectivity and data transfer. As an aside, it is instructive to note that the assumption $DU_P = DU_E$ and $(A_P)_P = (A_P)_E$, and a linear interpolation of p_e and p_w the "pressure smoothing" term in Eq. (13) turns out to be:

$$\frac{DU}{4A_P}(p_{EE} - 3p_E + 3p_P - p_W),\qquad(14)$$

which is, in essence, a third-order artificial dissipation term.

The pressure-correction equation arises upon decomposing the correct face velocity u_e into an estimated value, obtained with an approximate pressure field, and a corrective perturbation:

$$u_e = u_e^* + u_e' \qquad(15)$$

where $*$ denotes the approximate value and $u_e' \sim (DU/A_P)_e(p_P' - p_E')$, with the subscript 'e' in the RHS multiplier denoting a centred average of the values at the two neighbouring nodes on either side of the face. Substitution of Eq. (15) and analogous expressions for other face-velocity components into the continuity equation for the cell in Fig. 5, yields the following pressure-correction equation:

$$a_P p_P' = \sum_{m=E,W,N,S,T,B} a_m p_m' + R_m \qquad(16)$$

where

$$a_E = [\frac{\rho J^2}{A_P}q_{11}]_e,\ \ a_W = [\frac{\rho J^2}{A_P}q_{11}]_w,\ \ a_N = [\frac{\rho J^2}{A_P}q_{22}]_n,$$

$$a_S = [\frac{\rho J^2}{A_P}q_{22}]_s,\ \ a_T = [\frac{\rho J^2}{A_P}q_{33}]_t,\ \ a_B = [\frac{\rho J^2}{A_P}q_{33}]_b,$$

$$a_P = \sum_{m=E,W,N,S,T,B} a_m, \qquad(17)$$

with the mass imbalance R_m defined by:

$$R_m = U_w^* - U_e^* + V_s^* - V_n^* + W_b^* - W_t^* + [\frac{(\rho^o - \rho)J}{\Delta t}]_P \qquad(18)$$

The extension of the above pressure-correction scheme to transonic regime, including shock wave, involves two essential elements. First, the discretised momentum equations are modified to govern the flux variables, i.e. density-weighted velocities. The underlying idea is that variations in these quantities, particularly across shocks, tend to be far lower than variations in either density or velocity. As a consequence, numerical errors associated with inadequate resolution of gradients are reduced considerably. Second, a locally-adaptive, Mach-number dependent density retardation is introduced at nearly sonic and supersonic regions in order to give good shock-capturing resolution. This is achieved by modifying the contravariant velocities in the approximation of flow convection as follows:

$$U \longleftarrow \frac{\rho U}{\tilde{\rho}}, \quad V \longleftarrow \frac{\rho V}{\tilde{\rho}}, \quad W \longleftarrow \frac{\rho W}{\tilde{\rho}}, \tag{19}$$

where $\tilde{\rho}$ $(\equiv \rho - \bar{\mu}\vec{\rho}_x \Delta x)$ is the *retarded density* [see Hafez (1979)]. With attention focused on one-dimensional conditions, the convective flux $(\rho u \phi / \tilde{\rho})_x$ can be expressed as:

$$\left(\frac{\rho u \phi}{\tilde{\rho}}\right)_x = [u\phi(1 - \frac{\bar{\mu}\vec{\rho}_x \Delta x}{\rho})^{-1}]_x = (u\phi)_x + \underline{[\bar{\mu}(\frac{u\phi}{\rho})\vec{\rho}_x]_x} + HOT \tag{20}$$

where HOT denotes all higher-order terms,

$$\bar{\mu} = \text{MAX}\{0, \kappa[1 - (\frac{M_{ref}}{M})^2]\}, \tag{21}$$

and M_{ref} and κ are of the order O(1). With HOT discarded, the underlined term in (20) represents the dissipative mechanism in the present approach. For multi-dimensional flows, the retarded density $\tilde{\rho}$ is determined in the streamwise direction [see Lien & Leschziner (1993) for details].

2.2. MULTI-BLOCK ALGORITHM

The multi-block algorithm is based, essentially, on a sub-division of the solution domain into an arbitrary number of contiguous, non-overlapping blocks, each having its own grid and associated local coordinate system. Each grid is first generated separately by use of any suitable grid-generation procedure, the only constraint being continuity in grid-line positions across the block boundaries. Each block, looked in isolation, is then surrounded by an 'auxiliary' layer of two cells originating from neighbouring blocks. In effect, the block is made to penetrate into its neighbours to the extent of two cell intervals, in order to accommodate 'halo data' which are needed for the solution within the block in question. The choice of a two-cell penetration is linked to the nature of the higher-order convection scheme and the Rhie & Chow interpolation practice [recall Eqs. (9)-(14)]. Although the coordinate systems of neighbouring blocks can be quite different in orientation, as is exemplified in Fig. 7, all geometric data pertaining to the auxiliary layer attached to the parent block, including the metric tensors and the Jacobian, are treated in terms of the co-ordinate system of the parent block and stored as if the layer was part of the block. This arrangement obviates, with one exception noted below, the need for any one block to directly access the

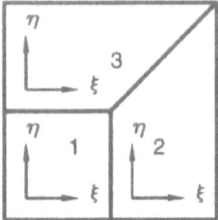

Figure 7. Multi-block arrangement with different local coordinate systems

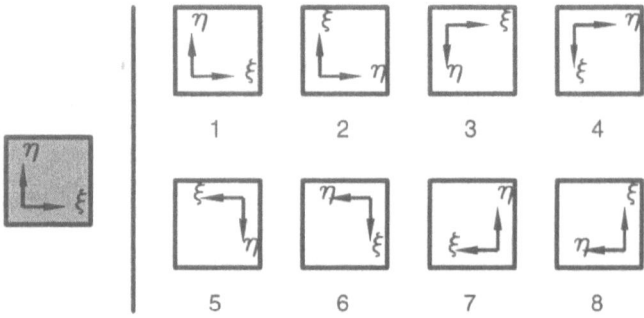

Figure 8. All possible permutations of local co-ordinate systems in blocks adjacent to any reference block

'foreign' geometric information and mass fluxes residing in neighbouring blocks during the solution process within the block.

Inter-block connectivity is handled by a connectivity matrix in the form of the 2D array MCONEC(BLOCK, FACE), where BLOCK is the block number being considered, FACE identifies the block face (ranging from 1 to 6, with 1 denoting the eastern face, 2 the western face, etc.) and MCONEC identifies the block sharing FACE with BLOCK. The coordinate system relating to any one block is stored in the form of COORD(BLOCK, FACE), representing all possible coordinate permutations in the neighbouring block sharing the face 'FACE'. To convey the basic idea without introducing a significant loss of generality, a typical 2D example for COORD is given in Fig. 8. The neighbouring RHS block can have any one of 8 combinations of coordinates, and this is signified by the integers 1 to 8. Another example illustrating the use of both MCONEC and COORD is given below by reference to Fig. 7, where:

$$MCONEC(3,1)=2, \quad MCONEC(3,2)=0,$$
$$MCONEC(3,3)=0, \quad MCONEC(3,4)=1,$$

and

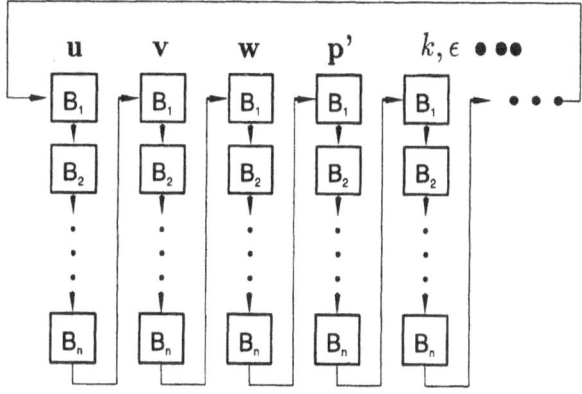

Figure 9.   Solution sequences of SIMPLE algorithm within a multi-block scheme

$$\text{COORD}(3,1)=8, \text{COORD}(3,2)=0,$$
$$\text{COORD}(3,3)=0, \text{COORD}(3,4)=1,$$

In the above, the value '0' signifies that the neighbouring block is a physical (real) boundary of the solution domain.

Reference to Eqs. (13) and (17) shows that the coefficients in the pressure-correction equation depend on the A_P value associated with momentum equations applied to the cell over which mass conservation is to be satisfied as well as to two neighbouring cells on either side in any coordinate direction. It is crucial, therefore, to transmit this quantity from the neighbouring blocks into the auxiliary two-cell layer when solving the momentum equations in the parent block. This transfer is greatly assisted by the fact that A_P is coordinate-invariant, i.e. independent of the block-local coordinate system.

Once the coefficients for the transport and pressure-correction equations have been assembled for each block, the resulting system of equations are solved in a segregated manner illustrated in Fig. 9. Each set of equations pertaining to any one block is solved within an 'inner iteration' by Stone's SIP or the ADI method concurrently with the temporarily 'frozen' block boundary conditions in the 'halo' region. Then, an update of boundary conditions is effected via the connectivity matrix and the identifiers of the coordinate systems in neighbouring blocks in order to establish the inter-block coupling. An 'outer iteration' consists of the solution of any one set of equations over all blocks and associated exchange of data across block boundaries. This sweep is arranged as a *Block Jacobi* method. Representative results obtained with the present multi-block scheme for flows over a 2D multi-element aerofoil (Lien et al, 1995c) and through a 3D linear turbine cascade (Lien & Leschziner, 1995a) are illustrated in Fig. 10.

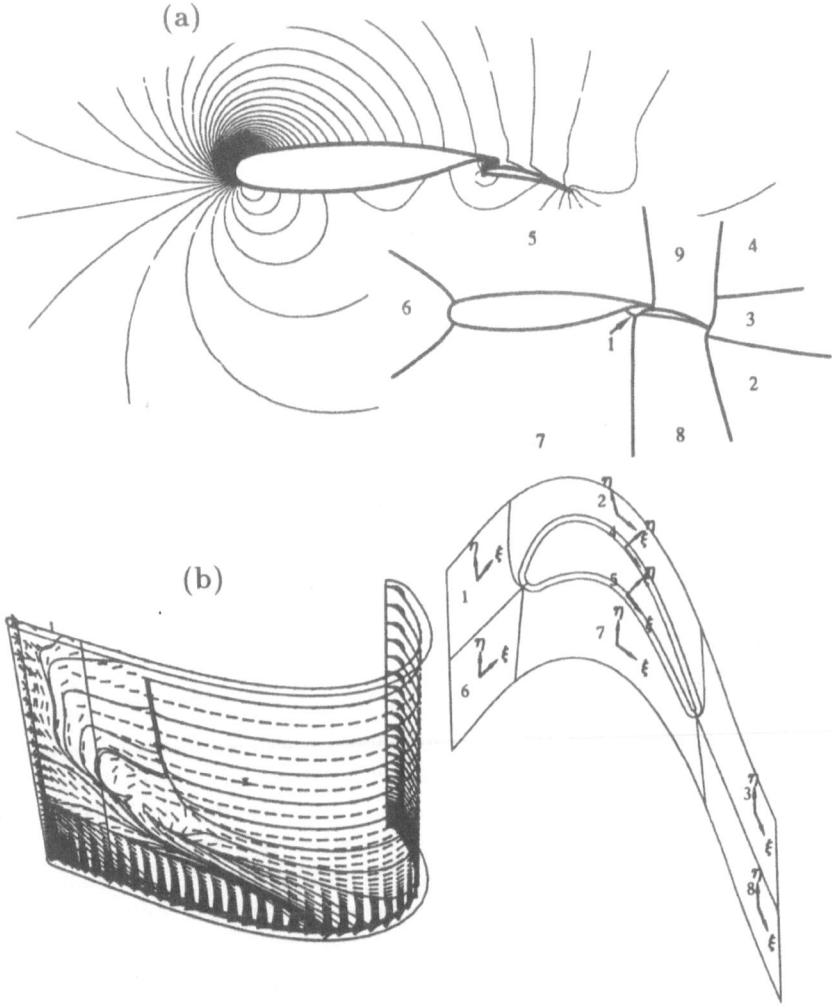

Figure 10. Application of multi-block scheme: (a) 2D multi-element aerofoil; (b) 3D linear turbine cascade

3. Turbulence Models and Their Numerical Implementation

3.1. NON-LINEAR EDDY-VISCOSITY MODEL

Conventional eddy-viscosity models are based on linear stress-strain relations - the Boussinesq relations. These are known to be the root of serious weaknesses in complex flows featuring curvature, swirl, separation, recirculation and impingement. A superior approach is one that is based on non-linear equations relating stresses to strain and vorticity components.

Based on series-expansion arguments, a general and coordinate invariant relationship between stresses and strains can be written as (Pope, 1975):

$$\frac{\overline{u_i' u_j'}}{k} = \frac{2}{3}\delta_{ij} - \frac{\nu_T}{k}S_{ij}$$

$$+C_1\frac{\nu_T}{\epsilon}[S_{ik}S_{kj} - \frac{1}{3}\delta_{ij}S_{kl}S_{kl}] + C_2\frac{\nu_T}{\epsilon}[\Omega_{ik}S_{kj} + \Omega_{jk}S_{ki}]$$

$$+C_3\frac{\nu_T}{\epsilon}[\Omega_{ik}\Omega_{jk} - \frac{1}{3}\delta_{ij}\Omega_{kl}\Omega_{kl}] + HOT$$

where C_μ and C_1 to C_3, proposed by Shih et al (1993) and applicable only to high-Re region, are

$$C_\mu = \frac{0.667}{A_1 + S + 0.9\Omega}|_{A_1=1.25}, \tag{22}$$

$$C_1 = \frac{3/4}{(1000 + S^3)C_\mu}, \quad C_2 = \frac{15/4}{(1000 + S^3)C_\mu}, \quad C_3 = \frac{19/4}{(1000 + S^3)C_\mu}, \tag{23}$$

and

$$S_{ij} = \frac{\partial u_i}{\partial x_j} + \frac{\partial u_j}{\partial x_i}, \quad \Omega_{ij} = \frac{\partial u_i}{\partial x_j} - \frac{\partial u_j}{\partial x_i}. \tag{24}$$

$$S = \frac{k}{\epsilon}\sqrt{\frac{1}{2}S_{ij}S_{ij}}, \quad \Omega = \frac{k}{\epsilon}\sqrt{\frac{1}{2}\Omega_{ij}\Omega_{ij}}. \tag{25}$$

To examine the effect of streamline curvature on turbulence, a third-order correction of Craft et al (1995) is also included, giving rise to

$$HOT = C_4\frac{\nu_T k}{\epsilon^2}[S_{ki}\Omega_{lj} + S_{kj}\Omega_{li} - \frac{2}{3}\delta_{ij}S_{km}\Omega_{lm}]S_{kl}$$

$$+ C_5\frac{\nu_T k}{\epsilon^2}(S_{kl}S_{kl} - \Omega_{kl}\Omega_{kl})S_{ij} \tag{26}$$

where $C_4 = 10C_\mu^2$ and $C_5 = -2C_\mu^2$.[3]

The turbulent viscosity ν_T, arising from $k - \epsilon$ modelling framework, is

$$\nu_T = C_\mu\frac{k^2}{\epsilon}. \tag{27}$$

In order to account for the semi-viscous near-wall effect, a damping function f_μ is introduced into (27), which, by reference to Norris/Reynolds' one-equation model (1975), can be derived as (Lien & Leschziner, 1995a):

$$f_\mu = [1 - \exp(-0.0198y^*)](1 + \frac{5.29}{y^*}), \tag{28}$$

---

[3]The original value for C_5 is $-5C_\mu^2$, which is found to be too strong to highly separated flow.

where $y^* = y\sqrt{k}/\nu$.

The dissipation rate ϵ is obtained from the solution of a related transport equation:

$$C_\epsilon - D_\epsilon = \frac{\epsilon}{k}(C_{\epsilon 1}P_k - C_{\epsilon 2}\epsilon), \qquad (29)$$

where

$$C_{\epsilon 1} = 1.44(1 + P'_k/P_k), \quad C_{\epsilon 2} = 1.92[1 - 0.3\exp(-R_T^2)] \qquad (30)$$

and $R_T = k^2/\nu\epsilon$. The P'_k is introduced to ensure that the correct level of near-wall turbulence-energy dissipation is returned. The end result is:

$$P'_k = 1.33[1 - 0.3\exp(-R_T^2)][P_k + 2\nu\frac{k}{y^2}]\exp(-0.00375y^{*2}) \qquad (31)$$

3.2. NUMERICAL IMPLEMENTATION

Following the algorithmic framework proposed by Lien & Leschziner (1995b) for second-moment closure, the numerical implementation of non-linear model also start with (4).

For the particular case $\phi = u$,

$$JS_u = -J\xi_x\frac{\partial}{\partial\xi}(\frac{p}{\rho}) - J\eta_x\frac{\partial}{\partial\eta}(\frac{p}{\rho}) - J\zeta_x\frac{\partial}{\partial\zeta}(\frac{p}{\rho})$$

$$-\frac{\partial}{\partial\xi}(J\xi_x\,\tau_{11}^{u\xi}) - \frac{\partial}{\partial\eta}(J\eta_x\,\tau_{11}^{u\eta}) - \frac{\partial}{\partial\zeta}(J\zeta_x\,\tau_{11}^{u\zeta})$$

$$-\frac{\partial}{\partial\xi}(J\xi_y\,\tau_{12}^{u\xi}) - \frac{\partial}{\partial\eta}(J\eta_y\,\tau_{12}^{u\eta}) - \frac{\partial}{\partial\zeta}(J\zeta_y\,\tau_{12}^{u\zeta})$$

$$-\frac{\partial}{\partial\xi}(J\xi_z\,\tau_{13}^{u\xi}) - \frac{\partial}{\partial\eta}(J\eta_z\,\tau_{13}^{u\eta}) - \frac{\partial}{\partial\zeta}(J\zeta_z\,\tau_{13}^{u\zeta})$$

$$(32)$$

where

$$\tau_{11}^{u\xi} = \overline{u'^2} + 2\nu_{eff}\xi_x(\frac{\partial u}{\partial\xi}), \quad \tau_{12}^{u\xi} = \overline{u'v'} + \nu_{eff}\xi_y(\frac{\partial u}{\partial\xi}), \quad \tau_{13}^{u\xi} = \overline{u'w'} + \nu_{eff}\xi_z(\frac{\partial u}{\partial\xi}),$$

$$(33)$$

and $\nu_{eff} = \nu_T + \nu$. Entirely analogous expressions relate to JS_ϕ for $\phi = v, w$.

To satisfy the 'no-slip' condition at wall, the *stress residuals* in (33) must be set, for a wall aligned with the $\xi - \zeta$ plane, as follows:

$$\tau_{11}^{u\eta} = 0; \quad \tau_{12}^{u\eta} = -\nu\eta_x v_\eta; \quad \tau_{13}^{u\eta} = -\nu\eta_x w_\eta.$$

$$(34)$$

Similar sets of expressions apply to the v- and w-equations.

Although additional arrays are required to store the Reynolds-stress tensors, the advantage of this particular implementation, which is compact and equally applicable to both eddy-viscosity models and second-moment closure, reduces significantly the complexity of the FORTRAN programming and, hence, promotes the opportunities for vectorisation [4]. This may be contrasted with the conventional approach, in which, even for the linear eddy-viscosity model, the expansion of the source term pertaining to the u-equation is:

$$JS_u = -J\xi_x \frac{\partial}{\partial \xi}(\frac{p}{\rho}) - J\eta_x \frac{\partial}{\partial \eta}(\frac{p}{\rho}) - J\zeta_x \frac{\partial}{\partial \zeta}(\frac{p}{\rho})$$

$$+\frac{\partial}{\partial \xi}[\mu_{eff}J(q_{12}u_\eta + q_{13}u_\zeta)] + \frac{\partial}{\partial \eta}[\mu_{eff}J(q_{12}u_\xi + q_{23}u_\zeta)] + \frac{\partial}{\partial \zeta}[\mu_{eff}J(q_{13}u_\xi + q_{23}u_\eta)]$$

$$+\frac{\partial}{\partial \xi}[\mu_{eff}J(c_1^\xi u_\xi + c_2^\xi u_\eta + c_3^\xi u_\zeta + c_4^\xi v_\xi + c_5^\xi v_\eta + c_6^\xi v_\zeta + c_7^\xi w_\xi + c_8^\xi w_\eta + c_9^\xi w_\zeta)]$$

$$+\frac{\partial}{\partial \eta}[\mu_{eff}J(c_1^\eta u_\xi + c_2^\eta u_\eta + c_3^\eta u_\zeta + c_4^\eta v_\xi + c_5^\eta v_\eta + c_6^\eta v_\zeta + c_7^\eta w_\xi + c_8^\eta w_\eta + c_9^\eta w_\zeta)]$$

$$+\frac{\partial}{\partial \zeta}[\mu_{eff}J(c_1^\zeta u_\xi + c_2^\zeta u_\eta + c_3^\zeta u_\zeta + c_4^\zeta v_\xi + c_5^\zeta v_\eta + c_6^\zeta v_\zeta + c_7^\zeta w_\xi + c_8^\zeta w_\eta + c_9^\zeta w_\zeta)]$$

with

$$q_{12} = \xi_x\eta_x + \xi_y\eta_y + \xi_z\eta_z, \quad q_{13} = \xi_x\zeta_x + \xi_y\zeta_y + \xi_z\zeta_z, \quad q_{23} = \eta_x\zeta_x + \eta_y\zeta_y + \eta_z\zeta_z.$$
$$(35)$$

and

$$\begin{bmatrix} c_1^\xi & c_2^\xi & c_3^\xi \\ c_4^\xi & c_5^\xi & c_6^\xi \\ c_7^\xi & c_8^\xi & c_9^\xi \end{bmatrix} = \begin{bmatrix} \xi_x\xi_x & \xi_x\eta_x & \xi_x\zeta_x \\ \xi_y\xi_x & \xi_y\eta_x & \xi_y\zeta_x \\ \xi_z\xi_x & \xi_z\eta_x & \xi_z\zeta_x \end{bmatrix}, \qquad (36)$$

$$\begin{bmatrix} c_1^\eta & c_2^\eta & c_3^\eta \\ c_4^\eta & c_5^\eta & c_6^\eta \\ c_7^\eta & c_8^\eta & c_9^\eta \end{bmatrix} = \begin{bmatrix} \eta_x\xi_x & \eta_x\eta_x & \eta_x\zeta_x \\ \eta_y\xi_x & \eta_y\eta_x & \eta_y\zeta_x \\ \eta_z\xi_x & \eta_z\eta_x & \eta_z\zeta_x \end{bmatrix}, \qquad (37)$$

$$\begin{bmatrix} c_1^\zeta & c_2^\zeta & c_3^\zeta \\ c_4^\zeta & c_5^\zeta & c_6^\zeta \\ c_7^\zeta & c_8^\zeta & c_9^\zeta \end{bmatrix} = \begin{bmatrix} \zeta_x\xi_x & \zeta_x\eta_x & \zeta_x\zeta_x \\ \zeta_y\xi_x & \zeta_y\eta_x & \zeta_y\zeta_x \\ \zeta_z\xi_x & \zeta_z\eta_x & \zeta_z\zeta_x \end{bmatrix}. \qquad (38)$$

[4]Generally, complexity may prevent or substantially inhibit vectorisation, because the required analysis is judged to be too demanding of system resource relative to anticipated performance improvement arising from the generated code.

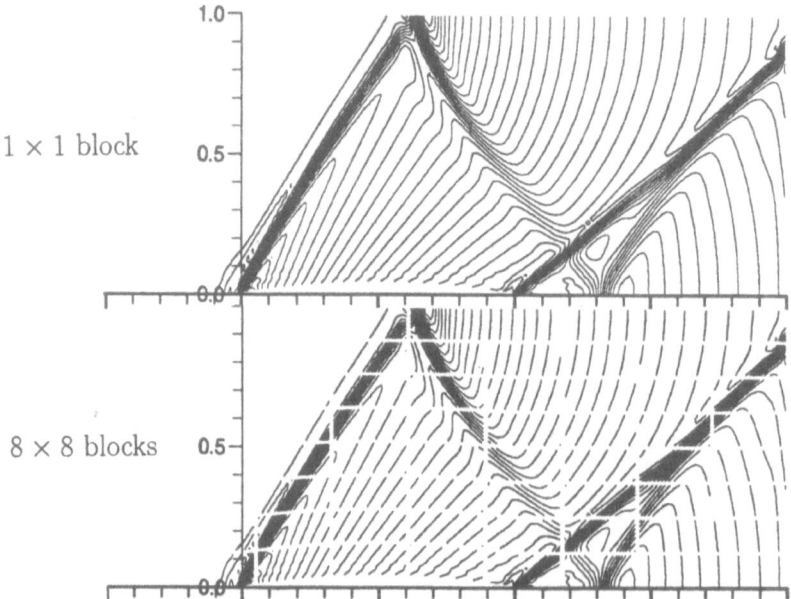

Figure 11. Supersonic bump: Mach-number contours obtained with single and multi-block (64-block) grids

4. Application

4.1. PORTING TO SHARED-MEMORY MACHINE

The first case investigated is a supersonic inviscid flow at M=1.4 over a 2D 4% circular bump, first considered by Ni (1982). Results were obtained over a grid of 128×64 nodes, of which 96×64 were uniformly distributed in the region downstream of the bump's leading edge. To verify that inter-block communication operates correctly, the results were obtained with single-block as well as multi-block grid arrangements. In the latter, each block contains 16×4 grid lines, giving rise to 64 blocks in total, which together form the single-block mesh. As demonstrated in Fig. 11, both results are identical. The shock reflections are well represented by the present pressure-correction method, and a comparison of performance with density-based methods, such as Lax-Wendroff and TVD schemes, can be found in Lien & Leschziner (1993).

The outline structure of the subroutine CALCU and the related flow chart of the computer code which is used to solve the u-momentum equation are shown in Fig. 12. Here, the intention has been to vectorise the innermost loop and parallelise the outer loop by use of *Autotasking* on CRAY Y-MP (8 CPUs). Autotasking is the compiling system, which, once initiated by *cf77 -Zp* command, can automatically determine those iterations of a

DO loop that are independent and then distribute loop iterations to the
various processors. The number of processors being used is controlled by
the environment variable NCPUS. Although wall-clock time can be reduced,
both CPU time and memory requirements will increase as a result of the
overhead incurred by the various programming steps associated specifically
with autotasking. Thus, CPU demands will increase through the insertion
of synchronisation points and the introduction of workload partitioning and
imbalance, while memory requirements will rise through the generation of
extra temporary variables and additional stack space. The speed-up ratio
on a dedicated system is defined as:

$$\text{speed-up ratio} = \frac{\text{wall-clock execution time (single CPU)}}{\text{wall-clock execution time (multiple CPUs)}}, \quad (39)$$

and the theoretical maximum speed-up according to Amdahl's law for parallelisation, analogous to that for vectorisation, is given by:

$$s_p = \frac{1}{(1 - f_p) + \frac{f_p}{P}}, \quad (40)$$

where

$$f_p = \text{fraction of task which can run in parallel}$$

$$P = \text{number of processors.}$$

As seen in Fig. 13, the speed-up ratio for the original programme, although 92% vectorised, is only 1.1 for 8 CPUs, corresponding approximately to a 10% level of parallelism. This ratio, which takes into account
Amdahl's law as well as the overhead associated with parallel processing,
was obtained, firstly, by executing the programme compiled with the -Zp
-Wu"-p" option and, secondly, by use of the *atexpert* utility. Profiling the
programme with *prof* and *profview* utilities showed that the LISOLV subroutine - the line-by-line TDMA solver - consumed the majority of CPU
time. This is because the TDMA is a recursive matrix-solving algorithm,
which allows the innermost loop to be vectorised, by use of the "Red &
Black" method, but still inhibits the parallelisation of the outer loop. In
order to enforce the parallelisation of the "DO NB" loop - NB being the
number of the block in question, this loop has to be taken out of the LISOLV
subroutine, i.e. LISOLV(ϕ) $\rightarrow$ LISOLV(ϕ,NB), followed by the insertion of
the CNCALL directive immediately preceding the loop. Similar considerations relate to the OVEL_Q subroutine, which is used to exchange the 'halo'
data after each relaxation sweep. By these means, both subroutines can be
parallelised as illustrated in Fig. 12, and the resulting speed-up ratios, also

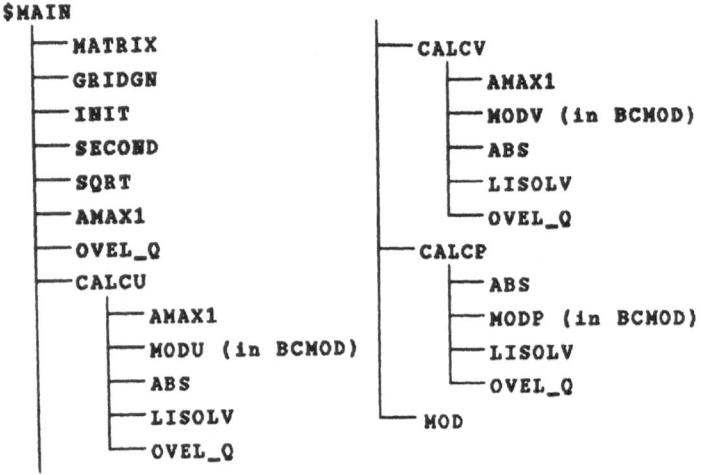

SUBROUTINE CALCU

▷ Assemble the coefficients A_m $(m = P, E, W, N, S)$
▷ Calculate the source term
▷ Modify the boundary conditions
▷ Calculate the residual

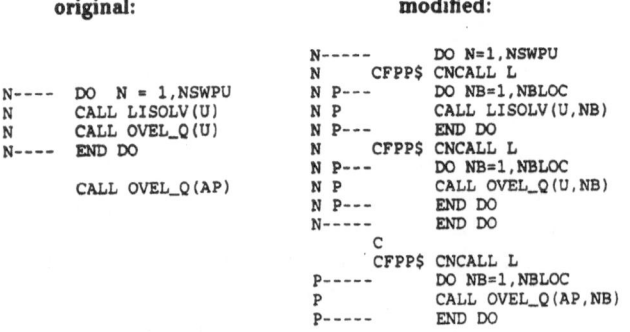

Figure 12. Supersonic bump: flow chart of the STREAM code and a partial FORTRAN listing of subroutine CALCU for the CRAY Y-MP

shown in Fig. 13, are 3.92 for 4 CPUs and 7.64 for 8 CPUs, corresponding approximately to 99% parallelism - a dramatic improvement over earlier performance figures. As OVEL_Q is the major fragment of the code, which is distinctly different for shared- and distributed-memory machines, a partial FORTRAN listing of this subroutine, pertaining only to the eastern 'halo' region, is given in Fig. 14. This will be contrasted in the next section with its counterpart for distributed-memory machines.

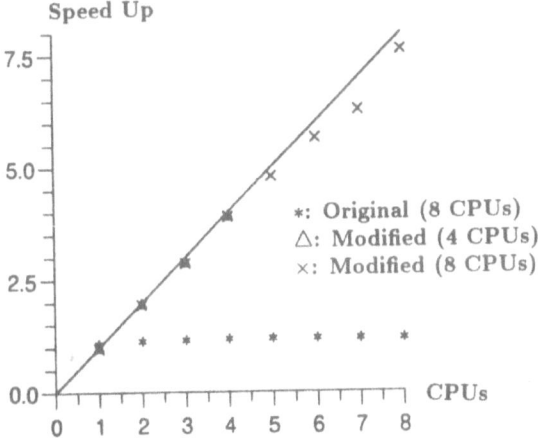

Figure 13. Supersonic bump: speed-up ratios executed on the CRAY Y-MP with Autotasking

```
CFPP$ NODEPCHK R
C
C East
C
        IF(MCONEC(NB,1).NE.0) THEN
V-      DO 101 J=NJBEG(NB)+2,NJEND(NB)-2
V       JJ=J-NJBEG(NB)+NJBEG(MCONEC(NB,1))
V       Q(NIEND(NB)-1,J)=Q(NIBEG(MCONEC(NB,1))+2,JJ)
V       Q(NIEND(NB)  ,J)=Q(NIBEG(MCONEC(NB,1))+3,JJ)
V- 101  CONTINUE
        END IF              Block 2
```

Q(NIEND(NB)-1,J)
Q(NIEND(NB) ,J)

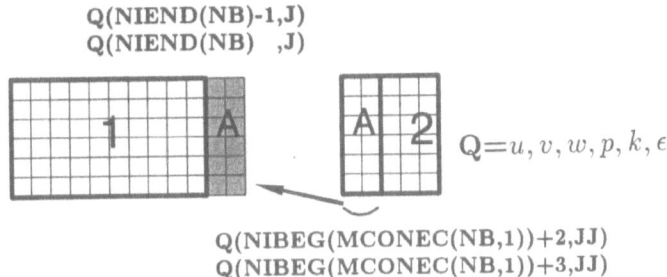

$Q = u, v, w, p, k, \epsilon$

Q(NIBEG(MCONEC(NB,1))+2,JJ)
Q(NIBEG(MCONEC(NB,1))+3,JJ)

Figure 14. Supersonic bump: a partial FORTRAN listing of subroutine OVEL_Q for the CRAY Y-MP

4.2. PORTING TO DISTRIBUTED-MEMORY MACHINE

In this section, two distributed-memory machines, namely the Intel iPSC/860 hypercube and CRAY T3D, were used to perform the computations reported later. The basic features of these two systems are summarised below:

- The Intel iPSC/860, installed at the Daresbury Laboratory in Cheshire, UK, consists of 2^d 40 MHz rx nodes hypercube, with dimension $d = 6$ such that each node is connected to d neighbours, the front-end processor (SRM) and a concurrent file system (CFS). The rx nodes, each of which contains 16 Mbytes of memory, use advanced pipelining techniques in software rather than vector hardware to achieve a large vector speed-up.

- The CRAY T3D, installed at the Edinburgh Parallel Computing Centre, consists of 160 nodes (or 320 PEs), arranged in a 3D torus. Each processing element (PE) consists of a DEC Alpha 21406 processor with 64 Mbytes of RAM, running at 150 MHz, supporting 64-bit integer and 64-bit IEEE floating point operations and delivering 150 64-bit Mflops/s. The Peak performance and aggregate memory of T3D itself are 48 Gflops/s and 20.48 Gbytes, respectively.

Calculations were first performed on the Intel iPSC/860 for the same test case as that detailed in Section 4.1 over a grid of 64 × 32 nodes, with the aim of optimising the code on a single node. As is evident from in Table 2, using the complier option -O4 -Mvect yields the best result. This option attempts to maximise the use of pipelining and dual-instruction mode. However, there is no hardware support for division within the loop, which inhibits pipelining. As a consequence, the -O4 -Mvect option is only 1.6 times faster than -O0 option (no optimisation).

TABLE 2. Supersonic bump: performance on the Intel iPSC/860 single node with different complier options

complier option	sec/per Δt	total time (sec)	$T/T_{[-O0]}$
-O0	0.7629	617.21	1.000
-O1	0.6309	510.42	0.827
-O2	0.5311	429.67	0.696
-O3	0.5182	419.29	0.679
-O3 -Mvect	0.4949	400.41	0.649
-O4 -Mvect	0.4893	395.86	0.641

Because the grid was partioned into 8 equal blocks, the workload is well balanced, as is demonstrated in Fig. 15 which was obtained by issuing the *ctool TX a.out* command, *a.out* being the executable file compiled with the *-O4 -Mvect -Mpref=comm* option. It is also seen in this figure that the communication overhead is approximately 10% of the overall computing time for a 128×64 grid.

To examine the effect arising from the adoption of synchronous and asynchronous message-passing routines on the efficiency of the solution procedure, two options were implemented in the OVEL_Q subroutines, the first using *csend, crecv* and the second *irecv, csend, msgwait*. Both options are given in Fig. 16. This is particularly informative if viewed in conjunction with Fig. 14. A distinct difference between the codes in Figs. 14 and 16 is that the latter requires explicit communication routines to exchange 'halo' data. The former, on the other hand, is fairly compact but requires, due to indirect addressing being involved, the insertion of the NODEPCHK directive in the subroutine to enforce vectorisation. The overheads arising from the synchronous and asynchronous communication routines shown in Fig. 16, including the packing and unpacking the messages, are 9.1% and 4.5%, respectively. This improvement is attributed to the use of overlapped communication techniques which allows data transfer to take place in parallel with programme execution, followed by *msgwait* to ensure the completion of such a transfer.

The parallel and total efficiencies are defined, respectively, as:

$$\text{parallel efficiency } \% = [\frac{\text{CPU time/per time step (single CPU)}}{\text{CPU time/per time step (multiple CPUs)}}]/P$$

$$(41)$$

$$\text{total efficiency } \% = [\frac{\text{total CPU time (single CPU)}}{\text{total CPU time (multiple CPUs)}}]/P \qquad (42)$$

where P is the number of processors. Values for both efficiencies achieved for the supersonic bump calculation using up to 16 processors over a 128×64 grid are given in Table 3. As seen, both parallel and total efficiencies decrease as the number of nodes increases, with the latter being slightly better than the former. Somewhat surprisingly, the total number of time steps required for convergence is also reduced as the the number of nodes increases. However, this observation only applies to the present flow and may not pertain to other circumstances. In general, convergence depends strongly and in a complex fashion on the choice of matrix solver, the construction of block topology and the complexity of flow physics.

Attention is next turned to the performance of the CRAY T3D for the same flow as that considered above, again using a 128×64 grid. Data is communicated either with the PVM message-passing library or with low-level shared-memory routines SHMEM. The T3D PVM implementation only allows Single Programme Multiple Data (SPMD) programming style.

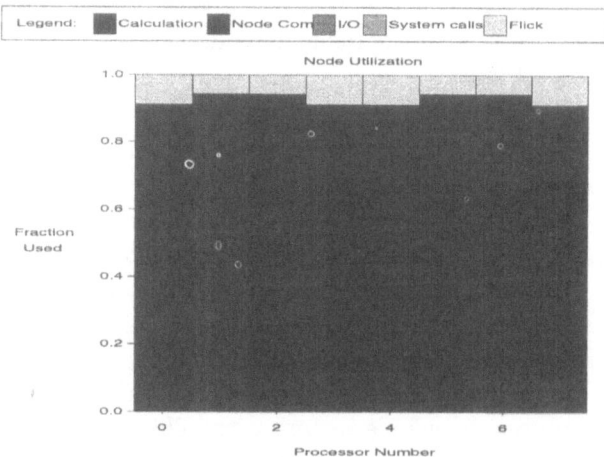

Figure 15. Supersonic bump: fraction of computation, communication, I/O, system calls and flick used on each Intel iPSC/860 node

TABLE 3. Supersonic bump: performance analysis on the Intel iPSC/860 with different number of nodes

No. of blocks	sec/per Δt	total CPU time (sec)	parallel efficiency (%)	total efficiency (%)
1×1	1.959	2875	100.0	100.0
2×1	1.000	1430	98.0	100.5
2×2	0.529	750	92.6	95.9
4×2	0.295	414	83.0	86.8
4×4	0.189	263	64.8	68.3

Thus, every PE in a partition[5] has a copy of the same executable file, and each of these processes is started at the same time when the *mppexec* command is issued. To improve the standard PVM 'send' and 'receive' functions, which are flexible in terms of data types but which also involve high latency and overhead penalties, PSEND/PRECV routines were included. These are effective means of passing short messages and default to the standard routines if the messages are long. The SHMEM routines are the most efficient, but care has to be exercised to avoid problems arising in relation

---

[5] A partition is a subset of PEs on the T3D, allocated as a unit for a particular programme to run.

Synchronous

```
C
C East
C
      IF(MCONEC(node+1,1).NE.0) THEN
C
C     packing message
C
          •

          •

      call csend(
     1MSGPTR(node+1)+1,QSDE,4*2*NY,MCONEC(node+1,1)-1,0)
C
      MGRIDE=crecv(
     1MSGPTR(MCONEC(node+1,1))+2,QREE,4*2*NY)
C
C     unpacking message
C
          •

C         •
      END IF
```

Asynchronous

```
C
C East
C
      IF(MCONEC(node+1,1).NE.0) THEN
      MGRIDE=irecv(
     1MSGPTR(MCONEC(node+1,1))+2,QREE,4*2*NY)
C
C     packing message
C
          •

          •

      call csend(
     1MSGPTR(node+1)+1,QSDE,4*2*NY,MCONEC(node+1,1)-1,0)
C
      END IF                              Block 2

      IF(MCONEC(node+1,1).NE.0) THEN
      call msgwait(MGRIDE)
C
C     unpacking message
C
          •

          •
C
      END IF
```

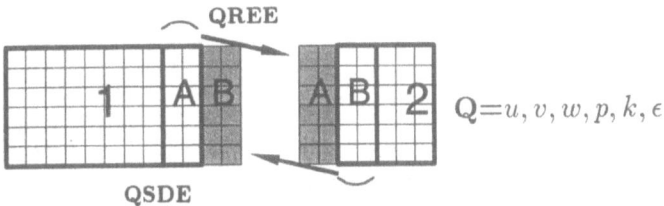

Figure 16. Supersonic bump: a partial FORTRAN listing of subroutine OVEL_Q for the Intel iPSC/860

to cache coherency. To ensure cache coherency, the 'remote write cache invalidate' bit should be set in all processors prior to any library routines being called. This bit is set by calling *shmem_set_cache_inv()* and cleared by calling *shmem_clear_cache_inv()*.

Prior to evaluating the performance associated with the PVM and SHMEM routines, the code was optimised on a single PE with different complier options. Related results are given in Table 4. As seen, the performance of *-O1 -Oscalar3* and *-O1* is identical: 5.65 times faster than *-O0* (no optimisation). In what follows, the *-O1 -Oscalar3* option was adopted, unless otherwise stated.

TABLE 4. Supersonic bump: performance on the T3D 1 PE with different complier options

complier option	sec/per iteration	$T/T_{[-O1-Oscalar3]}$
-O1 -Oscalar3	0.506	1.000
-O1	0.506	1.000
-O0	2.860	5.652

Table 5 contrasts the performance of PVM and SHMEM as alternative communication routines. To reduce the latency associated with the standard PVM send and receive functions, PSEND/PRECV routines were used; their latency levels are approximately three times shorter than the standard routines, as is shown in Table 6. Despite the substantial advantages offered by PSEND/PRECV, the parallel and total efficiencies achieved with PVM for 64 PEs are only 50.4% and 54.7%, respectively. By resorting to the SHMEM routines, these ratios increase dramatically to 78.0% and 84.1%, which is consistent with Table 6, where the latency for *shmem_put* and *shmem_get* is shown to be only $1 - 2$ μs. A fragment of FORTRAN code in the OVEL_Q subroutine, corresponding to that in Fig. 16 for the Intel iPSC/860 and containing the SHMEM routines, is presented in Fig. 17.

The performance on the T3D with SHMEM used for message passing, relative to that on the CRAY Y-MP (1 CPU), is indicated in Table 7. It is noted that the Y-MP is 14.3 times faster than a single PE on the T3D for a single-block solution. However, the ratio of $\frac{nT_n(T3D)}{T_1(Y-MP)}$ decreases as the number of blocks increases. This is due to a compensation effect arising, on the one hand, from a reduction in parallel efficiency in the numerator, which tends to increase the $nT_n(T3D)$ value, and on the other hand, from a degradation of vectorisation in the denominator, which can, as a result

TABLE 5. Supersonic bump: performance of PVM vs. SHMEM executed on the T3D with different number of PEs

PVM				
No. of blocks	sec/per Δt	total CPU time (sec)	parallel efficiency (%)	total efficiency (%)
1×1	0.4844	700	100.0	100.0
2×2	0.1277	179	94.8	97.8
4×4	0.0385	53	78.8	82.5
8×8	0.0150	20	50.4	54.7
SHMEM				
No. of blocks	sec/per Δt	total CPU time (sec)	parallel efficiency (%)	total efficiency (%)
1×1	0.4844	700	100.0	100.0
2×2	0.1270	171	99.3	102.3
4×4	0.0320	44	94.6	99.4
8×8	0.0097	13	78.0	84.1

TABLE 6. Performance of message-passing routines on the T3D

Library	Latency (μs)	Bandwidth (MB/s)
PVM	50-70	40-60
PSEND/PRECV	15-20	–
shmem_put	1-2	120
shmem_get	1-2	60

of shorter vector length, significantly increase the $T_1(Y - MP)$ value and hence decrease the ratio of the two.

The second case examined here is a three-dimensional subsonic turbulent flow over a 1:6 prolate spheroid at 30° incidence angle (Meier et al, 1984). The turbulence model adopted is that of Lien et al (1995a) described in Section 3.1, and results were obtained with 64^3 and 128^3 grids. The Reynolds number, based on the chord and free-stream velocity, is 6.5×10^6

```
C
C East
C
      IF(MCONEC(node+1,1).NE.0) THEN
C
C     packing message
C
           •

           •
C
      iput=shmem_put(
     1QREW,QSDE,2*NY,MCONEC(node+1,1)-1)
C
      END IF            Block 2
```

```
      IF(MCONEC(node+1,1).NE.0) THEN
      iget=shmem_get(
     1QREE,QSDW,2*NY,MCONEC(node+1,1)-1)
C                       Block 2
C     unpacking message
C

           •

           •
C
      END IF
```

Figure 17. Supersonic bump: a partial FORTRAN listing of subroutine OVEL_Q for the CRAY T3D

and the free-stream turbulence intensity is 0.2%.

An overview of the flow, including the surface streaklines and vortical structure following separation, is conveyed in Fig. 18. Azimuthal variations of skin-friction magnitude and direction are shown in Figs. 19 and 20, respectively, while circumferential distributions of wall pressure are given in Fig. 21. As seen at $x/2a=0.48$, the cubic variant of the non-linear $k-\epsilon$ model returns the best results, both in terms of the dip in the wall pressure and the peak in the skin-friction magnitude for $130° < \phi < 180°$. This is due to the ability of the cubic model, uniquely among the eddy-viscosity variants investigated, to account for the interaction between streamline curvature and turbulence - in this case resulting in an attenuation of shear stress on

TABLE 7. Supersonic bump: performance of the T3D vs. Y-MP (1 CPU)

no. of blocks	sec/per Δt T3D	sec/per Δt Y-MP	Mflops/s Y-MP	$\frac{nT_n(T3D)}{T_1(Y-MP)}$
1×1	0.4844	0.0338	170.3	14.3
2×2	0.1223	0.0394	148.6	12.4
4×4	0.0319	0.0478	126.1	10.7
8×8	0.0097	0.0775	82.7	8.0

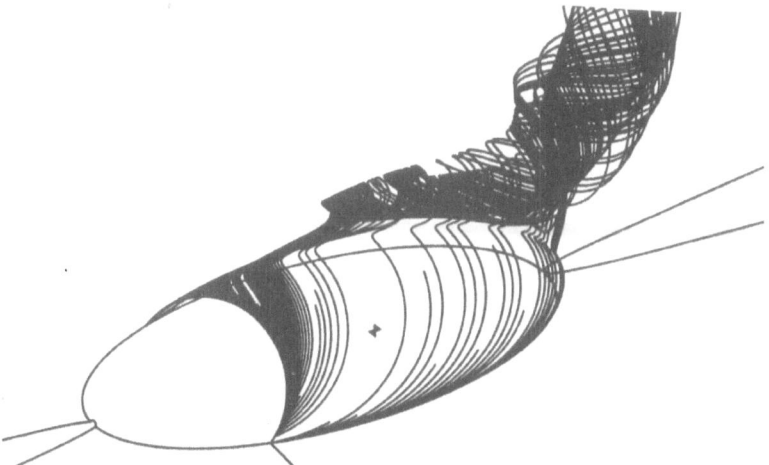

Figure 18. Subsonic ellipsoid: an overview of vortical structure detached from the separation lines close to the leeward side

the convex surfaces. Similar improvements achieved with the cubic variant is also observed at x/2a=0.73, in particular for $50° < \phi < 100°$, although all models fail to capture the free transition on the windward side.

Table 8 shows that computations were performed with three different compiler options, namely *-O1 -Oscalar3, -O1 and -O0*, for all variants of turbulence model indicated above. In the case of the linear model variant, the option *-O1 -Oscalar3* turns out to be the most efficient, and will be used later in all cases, unless otherwise stated. Also demonstrated in Table 8 is the fact that the extra computational costs associated with the quadratic and cubic variants, relative to the linear one, are only 3.8% and 9.1%, respectively. These penalties are much lower than that paid for using second-moment closure which yields a solution similar to those returned by

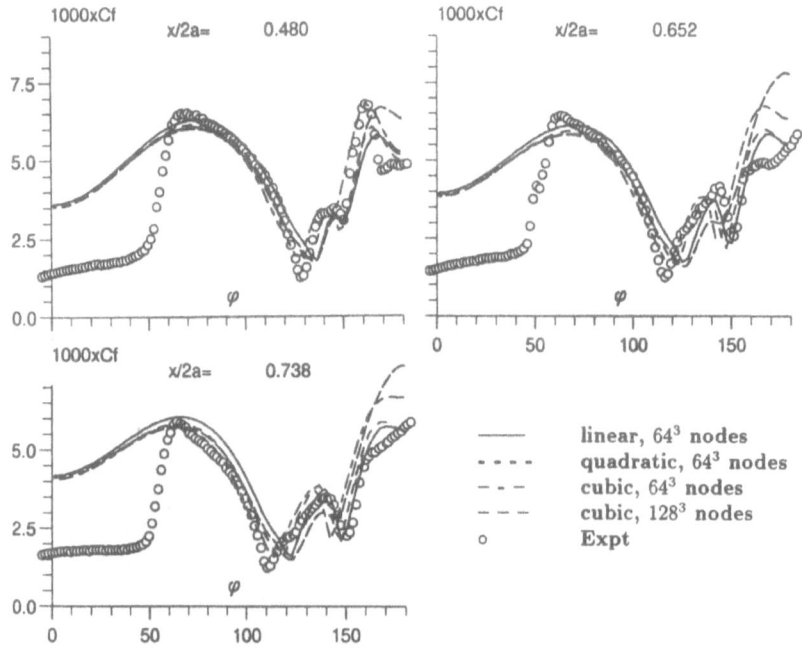

Figure 19. Subsonic ellipsoid: Azimuthal variations of skin-friction magnitude

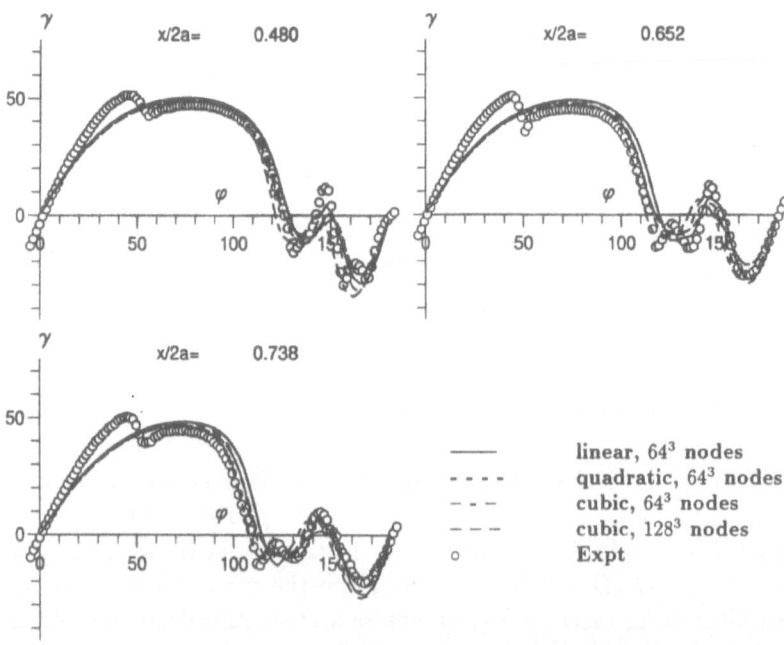

Figure 20. Subsonic ellipsoid: Azimuthal variations of skin-friction direction

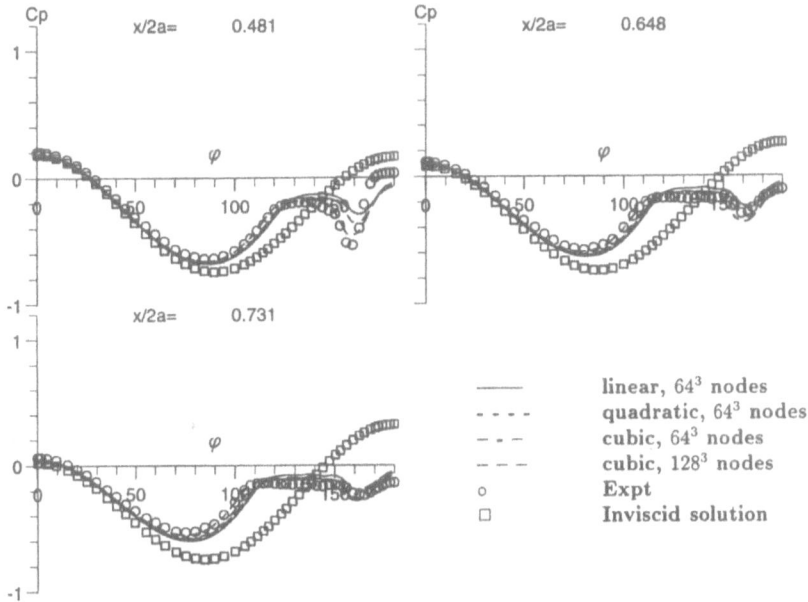

Figure 21. Subsonic ellipsoid: Azimuthal variations of pressure coefficient

the non-linear eddy-viscosity variants, but at CPU costs roughly $50 - 60\%$ in excess of that needed for the linear model [see Lien & Leschziner (1995c) for details].

TABLE 8. Subsonic ellipsoid: performance on the T3D 128 PEs with different complier options

model type	complier option	sec/per iteration	$T/T_{[-O1-Oscalar3]}$
linear	-O1 -Oscalar3	7.52	1.000
linear	-O1	7.64	1.016
linear	-O0	55.26	7.349
quadratic	-O1 -Oscalar3	7.81	1.038
cubic	-O1 -Oscalar3	8.21	1.091

The parallel efficiency achieved with the quadratic variant over a 64^3 grid is given in Table 9, and this performance is contrasted in Table 10 with that obtained with a 128^3 grid. Table 9 shows that the parallel efficiency expenditure per iteration also increases. However, as seen from Table 10, the increase for 64 PEs is only 7.3 ($<$ 8), and this ratio drops as the number of PEs increases, all signifying that communication overheads associated with the fine grid is lower than that with the coarse grid. In considering the implications of the figures contained in Table 10, it is finally important to point out that the number of iterations required to achieve convergence on the finer grid (subject to the same convergence criterion) increases from 1750 to 2100. This clearly has an effect on the figures in the last column of Table 10.

TABLE 9. Subsonic ellipsoid: parallel efficiency obtained with a 64^3 grid on the T3D

no. of blocks	sec/per iteration	parallel efficiency (T_8/nT_n, %)
8 ($4 \times 1 \times 2$)	15.48	100.00
16 ($4 \times 1 \times 4$)	7.72	100.25
32 ($4 \times 2 \times 4$)	3.80	101.83
64 ($8 \times 2 \times 4$)	2.12	91.28
128 ($8 \times 2 \times 8$)	1.08	89.56
256 ($8 \times 4 \times 8$)	0.56	86.38

TABLE 10. Subsonic ellipsoid: comparisons of CPU consumption obtained with 64^3 and 128^3 grids on the T3D

no. of blocks	sec/per itr. $N=64^3$	sec/per itr. $N=128^3$	$T(N=128^3)/T(N=64^3)$
64 ($8 \times 2 \times 4$)	2.12	15.48	7.30
128 ($8 \times 2 \times 8$)	1.08	7.81	7.23
256 ($8 \times 4 \times 8$)	0.56	3.84	6.86

5. Concluding Remarks

The formulation of a general multi-block algorithm for complex turbulent flows has been considered in detail. The algorithm couples, in an unstructured manner, finite-volume blocks in which the grid is structured but non-orthogonal, and in which the equations are solved on a fully-collocated grid by a pressure-correction algorithm.

The main impetus for developing a multi-block algorithm is derived from the need to accommodate geometric flexibility and the ability to implement the algorithm on parallel machines using a domain-decomposition approach. Treating each block as a sub-domain, with block-topology information stored in the connectivity matrices, allows the multi-block code to be ported to parallel architectures with relatively little effort.

As the number of block increases, the overall performance tends to decline. In the case of a shared-memory vector machine, this is mainly due to a shortening of vector length, while for distributed-memory machines, the decline in performance is attributable to the overheads arising from increased communication. On the Intel iPSC/860, overheads can be reduced by using an overlapped communication strategy, exemplified in Fig. 16. On the CRAY T3D, the most efficient way of reading and writing data is to use SHMEM routines, which, unlike PVM, are not portable and require special attention in relation to cache coherency.

The key to a further improvement of performance on the CRAY T3D appears to be a more effective use of the cache. A promising line of attack is to use the data in the cache as intensively and extensively as possible before data is replaced, to avoid cache-thrashing as a result of more than one operand in an expression maps of the same cache-line, and to maximise the usage of registers. However, applying the above strategy to a three-dimensional CFD code containing complex turbulence models is as much an art as it a science.

Acknowledgements

Part of this study documented herein have been supported by British Aerospace (CAL) and the Defence Research Agency in the UK. The computations were performed on the CRAY Y-MP at the Rutherford Appletion Laboratory, the Intel iPSC/860 at the Daresbury Laboratory and the CRAY T3D at the Edinburgh Parallel Computing Centre, the last via the CCAF programme, with allocations granted by the UK Science and Engineering Research Council. The author gratefully acknowledges this support. Thanks are also given to Professor M.A. Leschziner for his valuable comments on this manuscript.

References

Craft, T.J., Launder, B.E. and Suga, K. (1995) A Non-Linear Eddy-Viscosity Model Including Sensitivity to Stress Anisotropy, *Proc. 10th Symposium on Turbulent Shear Flows*, The Pennsylvania State University, **3**, 23.19.

Durbin, P.A. (1995) Constitutive Equation for the $k-\epsilon-\overline{v'^2}$ Model, *Proc. 6th Int. Symp. on Computational Fluid Dynamics*, Lake Tahoe, **1**, 258.

Gatski, T.B. and Speziale, C.G. (1993) On Explicit Algebraic Stress Models for Complex Turbulent Flows, *J. Fluid Mech.*, **254**, 59.

Hafez, M., South, J. and Murman, E. (1979) Artificial Compressibility Method for Numerical Solutions of Transonic Full Potential Equation, *AIAA J.*, **17**, 838.

Leonard, B.P. (1979) A Stable and Accurate Convective Modelling Procedure Based on Quadratic Upstream Interpolation, *Comp. Meth. Appl. Mech. Eng.*, **19**, 59.

Lien, F.S. and Leschziner, M.A. (1993) A Pressure-Velocity Solution Strategy for Compressible Flow and Its Application to Shock/Boundary-Layer Interaction Using Second-Moment Turbulence Closure, *ASME J. Fluid Engrg*, **115**, 717.

Lien, F.S. and Leschziner, M.A. (1994a) A General Non-Orthogonal Collocated FV Algorithm for Turbulent Flow at All Speeds Incorporating Second-Moment Closure, Part 1: Computational Implementation, *Comp. Meth. Appl. Mech. Eng.*, **114**, 123.

Lien, F.S. and Leschziner, M.A. (1994b) Upstream Monotonic Interpolation for Scalar Transport with Application in Complex Turbulent Flows, *Int. J. Num. Meth. Fluids*, **19**, 527.

Lien, F.S. and Leschziner, M.A. (1995a) Computational Modelling of a Transitional 3D Turbine-Cascade Flow Using a Modified Low-Re $k-\epsilon$ Model and a Multi-Block Scheme, *ASME Paper 95-CTP-80*.

Lien, F.S. and Leschziner, M.A. (1995b) Second-Moment Closure for Three-Dimensional Turbulent Flow Around and Within Complex Geometries, to appear in *Computers & Fluids*.

Lien, F.S. and Leschziner, M.A. (1995c) Computational Modelling of Multiple Vortical Separation From Streamlined Body at High Incidence, *Proc. 10th Symp. on Turbulent Shear Flows*, The Pennsylvania State University, **1**, 4.19.

Lien, F.S., Chen, W.L. and Leschziner, M.A. (1995a) Low-Reynolds-Number Eddy-Viscosity Modelling Based on Non-Linear Stress-Strain/Vorticity Relations, *Proc. 3rd Int. Symp. on Engineering Turbulence Modelling and Measurements*, May 27-29, 1996, Crete, Greece.

Lien, F.S., Chen, W.L. and Leschziner, M.A. (1995b) A Multi-Block Implementation of a Non-Orthogonal, Collocated Finite-Volume Algorithm for Complex Turbulent Flows, to appear in *Int. J. Num. Meth. Fluids*.

Lien, F.S., Chen, W.L. and Leschziner, M.A. (1995c) Computational Modelling of a High-Lift Aerofoils With Turbulence-Transport Models, *Proc. CEAS European Forum High Lift & Separation Control*, Bath, 10.1.

Meier, H.U., Kreplin, H.P., Landhauser, A. and Baumgarten, D. (1984) Mean Velocity Distribution in 3D Boundary Layers Developing on a 1:6 Prolate Spheroid With Artificial Transition, *DFVLR Report IB 222-84 A 11*.

Ni, R.H. (1982) A Multiple Grid Scheme for Solving the Euler Equation, *AIAA J.*, **20**, 1565.

Norris, L.H. and Reynolds, W.C. (1975) Turbulence Channel Flow With a Moving Wavy Boundary, *Report FM-10*, Dept. of Mech. Engrg., Stanford University.

Patankar, S.V. (1980) *Numerical Heat Transfer and Fluid Flow*, Hemisphere Publishing Corporation, London.

Pope, S.B. (1975) A More General Effective-Viscosity Hypothesis, *J. Fluid Mech.*, **72**, 331.

Rhie, C.M. and Chow, W.L. (1983) Numerical Study of the Turbulent Flow Past an Airfoil With Trailing Edge Separation, *AIAA J.*, **21**, 1525.

Shih, T.H., Zhu, J. and Lumley, J.L. (1993) A Realisable Reynolds Stress Algebraic Equation Model, *NASA TM-105993*.

Speziale, C.G. (1987) On Non-Linear $k-l$ and $k-\epsilon$ Models of Turbulence, *J. Fluid Mech.*, **178**, 459.

DATA-PARALLEL SOLUTION OF THE INCOMPRESSIBLE NAVIER-STOKES EQUATIONS

R.W.C.P. VERSTAPPEN AND A.E.P. VELDMAN
Department of Mathematics, University of Groningen
P.O. Box 800, 9700 AV Groningen, The Netherlands

1. Introduction

This contribution concerns the data-parallel solution of the incompressible, unsteady Navier-Stokes equations. We consider the following two applications: direct numerical simulation of turbulent flow (Sections 2 - 6) and sloshing of incompressible liquid inside moving containers (Section 7). The overlap of the two applications is fairly large. Among others, in both applications a finite-volume method is used, the velocities and pressures are defined on a staggered grid, the pressure gradient in the momentum equations and the incompressibility constraint are integrated implicitly in time; the convective and diffusive terms are treated explicitly. Nevertheless, the data-parallel methods that are used in the two applications are not the same. In particular, the methods for solving the pressure from a Poisson equation differ, and yet both methods are efficient within the context of their own application.

2. Direct numerical simulation of turbulence

Computer simulation has become a major tool for studying turbulent flow in many technological and environmental applications. It is not too much to say that computer simulation has changed the way in which cars, aircraft, or ships are designed. The turbulent flow in these applications typically involves a large range of dynamically significant scales of motion. Resolving the evolution of all these scales of motion from the unsteady 3D Navier-Stokes equations more than exhausts the presently largest computing resources. Thus, this approach - called direct numerical simulation (DNS) - is not feasible for car manufacturers, aircraft designers, ship builders, and

237

P. Wesseling (ed.), High Performance Computing in Fluid Dynamics, 237–260.
© *1996 Kluwer Academic Publishers.*

many others. For them acceptable computational effort can only be obtained by modelling the turbulence of the flow.

In current production codes the time-mean flow is resolved and the effect of any other motion on the mean is modelled. The next generation of codes will model less and resolve more; they will resolve the dynamics of the largest eddies and model the smaller eddies and the disordered motions. Using the fastest presently available computers and numerical algorithms, direct numerical simulation is limited to relatively low Reynolds numbers, i.e. to turbulent flows with a moderate range of scales of motion.

In many technological and environmental applications, the required simulation accuracy cannot be reached with the existing turbulence models: turbulence modelling forms the Achilles' heel of applied computational fluid dynamics. The detailed flow data that has been obtained using DNS plays a key role in improving our understanding of turbulent flow and in obtaining better turbulence models. Generally speaking, DNS provides much more detailed information about turbulence than experiments do.

High-performance computing is a prerequisite for direct numerical simulation of turbulence. Table 1 gives an overview of the requirements, in terms of processing power and memory size. The numbers in the column 'internal flow' are based on a DNS that has been performed (Verstappen and Veldman, 1996). The numbers in the other columns are based on estimates about the increase of the number of scales of motion as function of the Reynolds number. The numerical algorithms for the DNS of the wing and the car are taken to be as fast as today's algorithms for incompressible turbulent flows in rectangular geometries.

TABLE 1. Required memory and processing power for DNS

application	internal flow	wing	car
Reynolds number	10^5	10^6	10^7
number of grid points	10^7	3.10^9	10^{12}
floating-point computations	3.10^{14}	3.10^{17}	3.10^{20}
CPU performance (300 hours run)	300 Mflop/s	300 Gflop/s	300 Tflop/s
memory capacity	1 Gbyte	300 Gbyte	100 Tbyte
data transfer rate to main memory		100 Gbyte/s	100 Tbyte/s

About seven orders of magnitude have to be bridged to perform a DNS of a fully developed turbulent flow around a car, for example. Assuming that both computer hardware and computational algorithms will continue to progress at the rate that they have got ahead during the past three decades - both have become one and a half order of magnitude faster per decade - it will take roughly two decades to bridge the lacking orders of magnitude. For this estimate to come true, computers need to become three orders of magnitude faster and need to have three orders of magnitude more memory within the next two decades; within that span of time the numerical algorithms for DNS need to become three orders of magnitude faster, need to run efficiently at the fastest avaliable machines, and need to use three orders of magnitude less memory than todays algorithms do require. The latter is not often mentioned. Yet, it is most likely the hardest of all nuts to crack!

Today, already, their enormous appetite for memory bytes does restrict the size of direct numerical simulations. One way to overcome this restriction is to use a so-called domain-swapping technique. This technique requires machines with a hierarchy of memories. The entire flow data, containing all scales of turbulent motion at a certain time, is stored in a relatively slow background memory, of say 100 Tbyte. Then, part-by-part, the flow data is loaded in main memory, updated, and stored back. Thus the entire flow domain is updated for a short period of time, and this process is repeated untill the full evolution of the turbulent flow is computed. The last line in Table 1 shows the rate of data transfer from background to main memory and *vice versa* needed to swap two subdomains in a time equal to the time needed to update one subdomain, i.e. the rate to swap without slowing-down.

Presently, about $Re = 10^5$ is the highest attainable Reynolds number for a DNS. As an example, Figure 1 shows an averaged vorticity field in the symmetry plane of a 3D cubical lid-driven cavity at $Re = 50,000$, as obtained by DNS.

3. Computational procedure

To make this paper self-contained, the main lines of the method that has been used to compute the turbulent flow shown in Figure 1 are described concisely in this section. For a more detailed discussion of the computational procedure the reader is referred to Verstappen and Veldman (1994). The incompressible Navier-Stokes equations are discretized using a finite-volume method, where the velocities and pressures are defined on a staggered grid (cf. Harlow and Welsh (1965)). The pressure term and the incompressibility constraint are integrated implicitly in time; the convective and diffusive terms are treated explicitly. The computation of one time step can be divided into three sub-steps.

First, an auxiliary velocity $\breve{u}$ is computed by integrating the convective

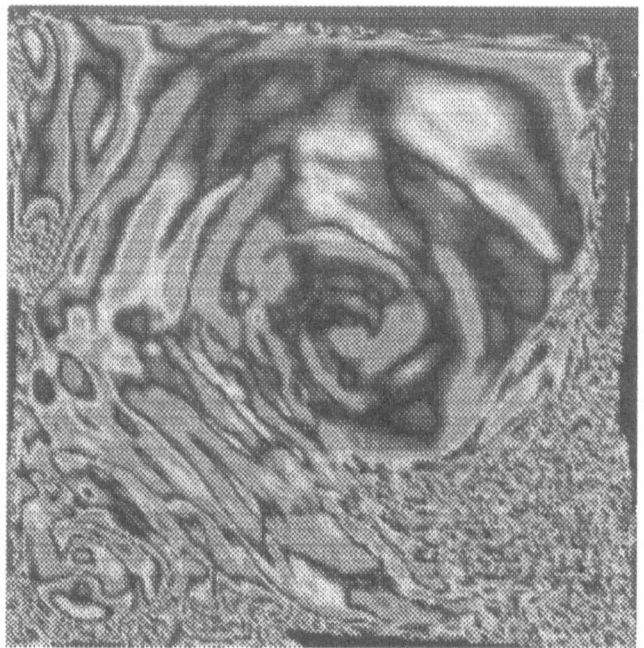

Figure 1.   An illustration of the structures in the symmetry plane of the cavity at Re=50,000. The orientation of the cavity is such that the upper-lid is driven from the left to the right. Shown is the third component of the vorticity vector in the symmetry plane of the cavity. Here, the vorticity is averaged over one large-eddy turn-around time.

and diffusive terms of the momentum equations over one time step δt

$$\tilde{u} \;=\; \frac{3u^n}{2} - \frac{u^{n-1}}{2}, \tag{1}$$

$$\breve{u} \;=\; u^n + \delta t \left(-\left(\tilde{u}\cdot\nabla_h\right)\tilde{u} + \frac{1}{Re}\Delta_h u^n \right). \tag{2}$$

Here, u^n denotes the (given) discrete velocity at time $t = n\delta t$. The spatial discretizations of the convective and diffusive term are represented by $\left(\tilde{u}\cdot\nabla_h\right)\tilde{u}$ and $\Delta_h u^n/Re$ respectively. The discretization of the convective term depends on both u^n and u^{n-1} since a second-order Adams-Bashforth method is used to integrate the convective term in time.

Next, the pressure p^{n+1} at time-level $t = (n+1)\delta t$ is computed from the Poisson equation

$$-\operatorname{div}_h\nabla_h p^{n+1} = -\operatorname{div}_h \breve{u}/\delta t. \tag{3}$$

And, finally, the divergence-free velocity u^{n+1} is obtained by adding the pressure term to the auxiliary velocity $\breve{u}$

$$u^{n+1} = \breve{u} - \delta t \nabla_h p^{n+1}. \tag{4}$$

The equations (1)-(4) hold in the interior of the spatial domain. At the boundaries Dirichlet, Neumann, mixed or periodic conditions can be prescribed, depending on the application.

4. Data-parallel implementation

Multiple processor computers can be configured as shared memory systems or as distributed memory systems. Shared memory system have one large, central memory that all processors can access directly. Memory accesses must be done in such a way that no two processors access the same data item at the same time. Generally this will slow down the speed of execution. On distributed memory systems each processor has its own memory. Data is distributed over all local memories. As before, mutual memory accesses must be excluded and all accesses must be synchronized. In addition, data items are shared only through explicit messages. The sending and receiving of data items takes time. To apply the full power of any distributed memory computer the communication time must be smaller than the time taken by the floating-point operations. Therefore, on a distributed memory system an efficient data-parallel implementation is less easy to achieve than on a shared memory system.

In Section 5 we will consider the data-parallel implementation on a shared memory system, namely a CRAY J932. Since the implementation of the explicit time-integration of the convection/diffusion equation (1)-(2) on the J932 is more or less trivial, and - more important - takes only a small part of the total computing time, it is undiscussed here. Solving the Poisson equation (3) takes most of the computing time on the J932 (no matter how it is done). Hence, this part should be implemented as efficient as possible. In Section 5 we will consider a Fast Fourier/Conjugate Gradient method for solving the discrete Poisson equation for the pressure.

After that, in Section 6, we will discuss the data-parallel implementation on a Connection Machine CM-5, a distributed memory system. Section 6 is divided into two parts. First (in Section 6.1) we will consider the parallelization of the sub-steps (1)-(2) and (4) of the time-marching procedure. The emphasis will be laid upon the time needed for the communications in relation to the time needed for the computations. In Section 6.2 data-parallel implementations of the Poisson solver on the CM-5 are discussed.

5. Parallelization on a shared memory machine: a CRAY J932

The CRAY J932 is a shared memory computer with 32 vectorprocessors. Each processor has a peak of 200 Mflop/s. Before applying multiple CPU's to a job, the individual performance of one (vector-)processor is to be optimized. For this reason we will discuss vectorization too.

Up to now, almost all direct numerical simulations have used periodic boundary conditions in the spanwise direction. This holds, for instance, for all the direct numerical simulations mentioned in the review paper by Reynolds (1990); for DNS of the turbulent boundary layer on a flate plate (cf. Spalart (1988)), for the direct numerical simulations of mixing layers (see for instance Moser and Rogers (1993)), for channel flows (e.g. Kim *et al.* (1987)), and so on. In general, flows that are statistically homogeneous in one spatial direction can be handled very well using periodic boundary conditions in that direction, provided that the domain is sufficiently large to contain all physically relevant modes in the statistically homogeneous direction.

Here we make explicitly use of the fact that the turbulent flow under consideration is statistically homogeneous in the spanwise direction. A uniform grid spacing in that direction is assumed. The Poisson equation for the pressure (3) is solved using a combination of a Fast Fourier Transform method in the spanwise direction and an Incomplete Choleski Conjugate Gradient method in the spectral space.

The discrete Poisson equation for the pressure $p(i, j, k)$ reads

$$\mathbf{L}p(i, j, k) - \frac{p(i, j, k - 1) - 2 * p(i, j, k) + p(i, j, k + 1)}{\delta z^2} = -r(i, j, k) \quad (5)$$

where $i = 1, .., nx$, $j = 1, .., ny$, and $k = 0, .., nz - 1$; $r(i, j, k)$ denotes the right-hand side of the Poisson equation, δz is the spacing in the spanwise direction k, and $\mathbf{L}p(i, j, k)$ is equal to

$$lx(i) * p(i - 1, j, k) \quad + \quad lx(i + 1) * p(i + 1, j, k) \quad +$$
$$ly(j) * p(i, j - 1, k) \quad + \quad ly(j + 1) * p(i, j + 1, k) \quad + \quad d(i, j) * p(i, j, k)$$

where the coefficients lx, ly and d depend on the grid spacing in i and j direction. The diagonal d is strictly positive. After a Fourier transformation in the statistically homogeneous direction k, the discrete Poisson equation (5) falls apart into a set of mutually independent equations of the form

$$\left(\mathbf{L} + 2 \cos \frac{2\pi m}{nz} \mathbf{I} \right) \hat{p}(i, j, m) = \hat{r}(i, j, m) \quad (6)$$

where $m = 0, 1, .., nz/2$. The complex quantities $\hat{p}$ and $\hat{r}$ are the Fourier transforms of p and r respectively:

$$\hat{p}(i,j,m) = \sum_{k=0}^{nz-1} p(i,j,k)e^{\frac{2\pi i k m}{nz}} \quad \text{and} \quad \hat{r}(i,j,m) = \sum_{k=0}^{nz-1} r(i,j,k)e^{\frac{2\pi i k m}{nz}}$$

where $m = 0, 1, .., nz/2$.

For simplicity we assume that nz is a power of 2. The above transforms can then be computed using the standard Fast Fourier Transform method. The j-loop in the calculation of the FFT's can be divided in equal chunks that are passed to the processors; the inner-i-loop of each chunk can be executed in vector mode. For the optimization of the transform itself the reader is referred to Van Loan (1992). Thus an optimal parallel-vector implementation is obtained. The vector length (nx) and the number of chunks (ny) are typically of the order of several hundreds. So both are (more than) sufficiently large for the CRAY J932 to run at high speed. Obviously, the scaling is perfect: the CRAY tool for analyzing the parallelism in the FFT-code predicts a scaling of 31.7 out of 32.

Once the FFT's are computed the equations (6) can be solved in parallel. These equations are in fact Poisson equations in two spatial dimensions (indexed with i and j respectively) with additions to their diagonals due to the Fourier transforms. It may be emphasized that the additions have the correct sign: the diagonals become stronger.

The set of 2D Poisson equations plus diagonal are solved in parallel using an Incomplete Choleski Conjugate Gradient method for each equation. Since the equations (6) are mutually independent the parallelization is straightforward, and scales 100%:

FOR ALL m DO PARALLEL solve the 2D Poisson equations (6)

The vectorization of the ICCG method, and the distribution of the work will be discussed next.

We start with the vectorization of ICCG. The preconditioner is constructed from an incomplete Choleski decomposition without fill-in; see Meijerink and Van der Vorst (1977). This decomposition is modified according to Gustafson (1978). The preconditioner is time-independent and is computed only once, namely before the time stepping starts. Consequently, the time needed to construct the preconditioner is insignificant. By using Eisenstat's implementation, the preconditioned system can be solved iteratively for practically the same cost as the unpreconditioned system (Eisenstat, 1981): the preconditioned Poisson system requires 22 floating point operations per grid point and iteration; the unpreconditioned CG-iteration requires 19. Thus, in terms of floating point operations per iteration the preconditioner comes almost for free. Yet, in terms of CPU-time per iteration

the preconditioned CG iteration is more expensive than the uncondi-
tioned one. This due to the fact that the floating point operations of the
ICCG are done at a lower speed.

In two spatial dimensions, the incomplete Choleski preconditioner introduces a recursion in both directions over the grid. A typical recursive
relation is of the form:

$$p(i,j) \ = \ r(i,j) \ - \ a(i,j) * p(i-1,j) \ - \ b(i,j) * p(i,j-1)$$

See also Figure 2. The element $p(i,j)$ depends on its previously computed

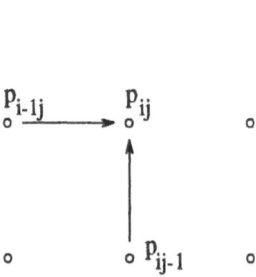

Figure 2.   The data-flow introduced by the incomplete Choleski preconditioner.

neighbors in i and j direction. However, the elements $p(i,j)$ on a diagonal
$i+j$ = constant depend only on values of p corresponding to a previous diagonal, and thus, in a diagonal-wise ordering, the unknowns can be computed
in vector mode. This observation is explored on vector-computers: the vectors correspond to diagonals of the grid; see e.g. Dongarra *et al.* (1991). We
have re-ordered the unknowns explicitly in a diagonal-wise manner. The
re-ordering is illustrated in Figure 3.

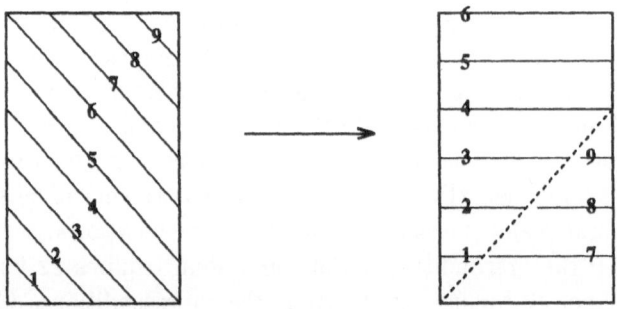

Figure 3.   Explicit diagonal re-ordering in two spatial dimensions.

All the arrays that are involved in the ICCG iteration are explicitly overwritten in this way. It may be noted that the re-ordering can be performed in space, i.e. without additional storage. The average vector length is equal to the average length of the diagonals, which is either of the order of nx or of the order of ny. In practice, these vectors are (more than) sufficiently long.

Table 2 gives some typical values for the performance of the explicitly diagonal re-ordered ICCG method and - for comparison - for the unpreconditioned CG method for a Poisson equation. The use of the preconditioner

TABLE 2. Performance of CG and ICCG on two vectorcomputers

	CRAY J90 (one processor) 200 Mflop/s peak		NEC SX-3 (one processor) 2750 Mflop/s peak	
	Mflop/s	% peak	Mflop/s	% peak
CG	140	70%	1100	40%
ICCG	80	40%	700	25%

reduces the number of iterations needed to converge. For a 2D Poisson equation (without an addition to its diagonal) the net gain of the use of the preconditioner on a vectorcomputer is about a factor of three. Here, we have counted for the reduction of the number of iterations, the reduction of computational speed, and the slight increase of floating point operations per iteration.

So far for the vectorization of an ICCG iteration for a Poisson equation in two spatial dimensions. The equations (6) have an additional weight on their diagonal due to the FFT. This weight increases with the frequency in the statistically homogeneous direction. Hence, the ICCG iterations for the higher frequencies converge much faster than the iterations for the lower frequencies do. As a result of this, an even distribution of the equations over the processors can lead to an uneven distribution of the work over the processors. In general this point deserves careful attention, and compiler directives are to be used on the CRAY J932 to distribute the work in an even manner. In the present application, however, the lowering of the number of iterations with the frequency is almost completely undone by the worsening of the initial guess with the period. The initial guesses for the ICCG iterations are obtained by extrapolating the pressures from previous time-levels. Therefore, the initial guesses for the lower frequencies are much more accurate than the initial guesses for the higher frequencies are. All things considered the net differences in work per equation are too small to bother about.

The FFT/ICCG Poisson solver runs at about 80 Mflop/s on one vectorprocessor of the CRAY, and scales up perfectly. Thus 2.5 Gflop/s can be achieved on 32 processors.

6. Parallelization on a distributed memory machine: a Connection Machine CM-5

The Connection Machine CM-5 as installed at the University of Groningen is a 16 processor node system with a peak performance of approximately 2 Gflop/s. Each processing node is a 128 Mflop/s computational unit composed of a SPARC micro-processor, 4 vector units, 32 Mbytes of memory and a network interface. The structure of the data network is a so-called fat tree. The data network guarantees 10 Mbyte/s to each processing node, no matter where in the system the data is being sent.

The CM-5 can be programmed both on a global level (using data parallel Fortran or data parallel C) and on a local level (in a message-passing programming style). We restrict ourselves to the global level of programming and use CM-Fortran, which is practically identical to the language Fortran 90. Large parts of the CM-Fortran code have been generated from an existing Fortran 77 program by using CMax, which is a CM-5 software tool that automatically converts Fortran 77 programs to CM-Fortran.

6.1. TIME-INTEGRATION OF CONVECTION/DIFFUSION ON THE CM-5

The integration of the convection/diffusion equation (1)-(2) forms no problem on a shared memory system like a CRAY J932. On a distributed memory system, however, the computation of (1)-(2) induces communication between the processors. In this section, we will investigate the costs of this.

The computation of the sub-steps (1)-(2) and (4) of the explicit time-marching procedure can be done in parallel by letting each processor treat its own subdomain. In the sequel, we will focus on the data-parallel execution of (2); the parallelization of (4) goes along the same lines; the parallelization of (1) is trivial. We kick off by discussing various ways to compute (2) data-parallel. Next, we will analyze the fastest way by comparing the times for communication and computation.

We have started by simply converting our existing Fortran 77 code into a CM-Fortran code using CMax. The thus obtained data-parallel Fortran code performed rather disappointingly: its megaflop-rate lies within the range of a $5,000 workstation. The reason for this is that the staggered location of the components of the velocity and the pressure is not recognized by the software. All DO-loops are simply replaced by FORALL statements. This CM-Fortran construct is much slower than other constructs that can do the same job, a WHERE statement, or a MERGE statement, e.g. Before discussing these faster constructs, we will consider the data-structure in detail.

The flow domain is divided into finite volumes. The discrete pressure is defined at the centre of each volume; the discrete velocity components are defined at the cell-faces, namely such that the velocity component perpendicular to a cell-face is defined in the middle of that cell-face. The staggering of the grid is sketched in Figure 4 (in two spatial dimensions).

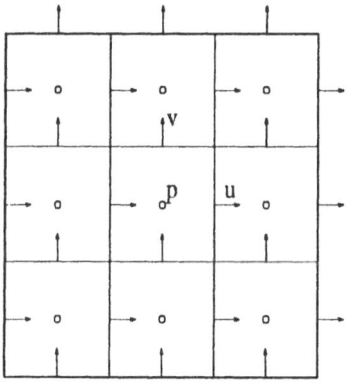

Figure 4. The location of the discretized velocity (u,v) and pressure p on the staggered grid (in two spatial dimensions).

The computation of velocity components at the boundaries differs from those at internal grid points. The velocities at internal grid points are to be updated according to (2); at the boundaries different expressions hold. This can be realized by a WHERE statement of the following form

WHERE (''not at the boundary'')
$$\breve{u} = u^n + \delta t \left(- \left(\tilde{u} \cdot \nabla_h \right) \tilde{u} + \frac{1}{Re} \Delta_h u^n \right)$$

The condition "not at the boundary" differs for the three components of the velocity, due to the staggering of the grid. Thus, three masks are to be constructed: for each component of the velocity one. These masks are independent of time, i.e. need to be computed only once, provided that there is enough space available to store them during the whole time-integration. Instead of using a WHERE statement, the update of the components of the velocity can be computed using array-sections, or alternatively, the update can be done unconditionally, followed by a reparation of the violated boundary conditions. All these three solutions are significantly faster than CMax's solution, i.e. than a FORALL statement.

We have compared all above mentioned solutions and found that the unconditional update followed by a reparation of the conditions at the boundaries is the fastest and uses the fewest memory. It is approximately twice as fast as the solution using a WHERE statement, and it is more than an order of magnitude faster than the FORALL solution.

We have implemented the fastest solution. Hereto, another consequence of the staggering of the grid has to be considered. Namely that the three arrays - say u, v and w - containing the three components of the discrete velocity and the array p of discrete pressures are not conformable, i.e. their dimensions differ. Indeed, take nx volumes in the first spatial direction (the velocity component in this direction is denoted by u), ny volumes in second direction (velocity component v) and nz volumes in the third direction (w).

Then, the dimensions of the arrays u, v, w and p become

$$u(0:nx, 1:ny, 1:nz) \qquad v(1:nx, 0:ny, 1:nz)$$

$$w(1:nx, 1:ny, 0:nz) \qquad p(1:nx, 1:ny, 1:nz)$$

Adding two non-conformable arrays, for instance u and v, makes no sense in CM-Fortran (nor in Fortran 90). This also holds for other operations. Therefore, all four arrays u, v, w and p are redefined such that they become conformable. That is, all dimensions are taken equal to nx*ny*nz. The "missing" elements, which correspond to prescribed velocities at the boundaries, are stored separately. Then, all updates can be performed unconditionally, i.e. for i=1,..,nx, j=1,..,ny and k=1,..,nz, and the thus violated boundary conditions can be repaired afterwards. It may be noted that this solution is rather laborious for the programmer, since it involves a change of the data structure.

To estimate the ratio between the time needed for communications and time taken by the computations, we will count the number of shifts and floating-point operations needed to integrate the convective/diffusive part of the Navier-Stokes equations over one time step.

Shifting is an intrinsic operation of CM-Fortran (and also of Fortran 90). In fact there are two types of shift, called CSHIFT and EOSHIFT. The "C" in CSHIFT stands for circular; "EO" means end-off. The EOSHIFT allows one to incorporate Dirichlet boundary conditions; the CSHIFT assumes periodicity in the direction of the shift. For instance, let a(1:nx,1:ny) and b(1:nx,1:ny) be two-dimensional, conformable arrays, then the statement

$$b = \text{CSHIFT}(a, \text{DIM} = 2, \text{SHIFT} = 1)$$

causes the elements of **b** to become equal to

$$b(i, j) = a(i, j + 1)$$

for i=1,...,nx and j=1,..,ny-1 and

$$b(i, ny) = a(i, 1)$$

for i=1,...,nx. The integration of the discretized convective/diffusive terms of the Navier-Stokes equations requires nearest-neighbor shifts only. Figure 5 shows the elements that are involved in the update of an element u_{ij}. A similar figure can be drawn in three spatial dimensions, and for the other components of the velocity vector.

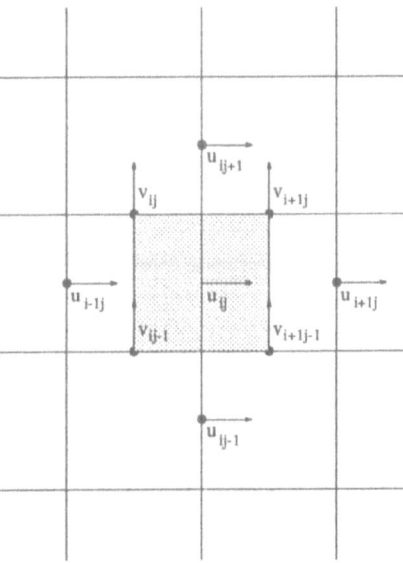

Figure 5. Array elements of u and v needed to update u_{ij}.

The second-order central discretization of the diffusive part of the two-dimensional Navier-Stokes equation results into the well-known five-point molecule. Its evaluation requires four shifts (by plus and minus one in the first direction, and by plus and minus one in the second direction). On a stretched grid, all five elements of the stencil have to be multiplied by different constants and have to be summed together. This costs nine floating-point operations. Thus, in two spatial dimensions, the ratio between shifts and floating-point operations for the diffusive part is approximately equal to 0.44. This ratio can also be determined for the evaluation of the convective term of the Navier-Stokes equation. The results are summarized in Tables 3 and 4, for two and three spatial dimensions, respectively.

TABLE 3. Ratio of shifts and flops in two dimensions

2D	convection	diffusion	convection & diffusion
shifts	7	4	11
flops	28	9	38
ratio	0.25	0.44	**0.29**

TABLE 4. Ratio of shifts and flops in three dimensions

3D	convection	diffusion	convection & diffusion
shifts	12	6	18
flops	44	13	58
ratio	0.27	0.46	**0.31**

N.B. In Tables 3 and 4, the number of shifts and flops is counted per equation. Note that there are two momentum equations, for both components of the velocity one, in 2D and three in 3D.

As can be seen from Tables 3 and 4, the ratio between (nearest-neighbor) shifts and floating point operations, i.e. the ratio between communication and computation, for one evaluation of a second-order finite-volume discretization of a convection-diffusion equation is approximately 0.3 in both two and three spatial dimensions.

From this ratio, the communication costs can be estimated. To obtain an estimation of the communication costs, we will count the number of data-elements that have to be moved from a local memory to another for one shift first. Here, we consider the operation CSHIFT(u,DIM=1,SHIFT=-1) on a target machine with p local memories. The array u has N entries. It can either be a two or a three dimensional array. In two dimensions it is defined as u(1:N_2,1:N_2), where $N_2^2 = N$; in 3D we take u(1:N_3,1:N_3,1:N_3), with $N_3^3 = N$. To ease the counting, we assume that the 2D array u can be divided into p subarrays of size $(N_2/p^{1/2})^2$, and that all elements of one subarray are stored in one local memory. Likewise, we assume that the 3D array u consists of p subarrays of size $(N_3/p^{1/3})^3$, and that the elements of one subarray are located in one local memory. Then, the absolute number of data-motions is equal to $(N_2/p^{1/2})*p$ in 2D and $(N_3/p^{1/3})^2*p$ in 3D. The relative number of data-motions are $(p/N)^{1/2}$ and $(p/N)^{1/3}$ respectively.

Each node of a 16 node CM5 has 4 vector units, and each vector unit has its own local memory. Thus, in total, there are 64 local memories. The 16 nodes can communicate at a speed of 10 Mb/s. Two local memories within one node communicate at 20 Mb/s. Hence, the average speed of communication is $0.25 * 10 + 0.75 * 20 = 17.5$ Mb/s. We take 10^6 data-elements of 8 bytes each. Then, the total number of bytes to be moved for the evaluation of one convection-diffusion equation in two spatial dimensions can be estimated by $8 \cdot 10^3 * 11 * 8 = 704000$. This data-motion takes approximately 0.04 seconds (at a speed of 17.5 Mb/s). We have measured the actual time that the shifts take, and found that the actual time equals the estimation: both come to 0.04 seconds per equation.

As remarked before, the ratio between shifts and flops is approximately equal to 0.3. Now, let us assume that the flops are for free, i.e. that they can be overlapped with the communications. Then, $38 \cdot 10^6$ floating point operations would take 0.04 seconds, and the time-integration of the 2D convection-diffusion equation would run at 950 Mflop/s (46% of peak). This thought experiment shows that the communication slows down the performance. It goes without saying that the actual Mflop rate has been measured: the time-integration of the convection-diffusion equation on a 1000^2 grid runs at approximately 15% of the peak. Thus, the ratio between the communication time and computation time equals 1 to 2.

In three spatial dimensions using a 100^3 grid, we have measured a time of 0.6 seconds for the shifts required to evaluate three convection-diffusion equations. This limits the speed to 300 Mflop/s (15% of peak), where the maximum can only be reached if the flops are fully overlapped with the communications. The latter is not the case: the actual speed is approximately 7% of the peak, i.e. the communication time equals the time of the computations.

We conclude this section, by summarizing the main results in Table 5.

TABLE 5. A convection/diffusion equation on the CM-5

| | speed | | computation : communication |
	Mflops	% peak	
2D	300	15%	2:1
3D	150	7%	1:1

6.2. HOW TO SOLVE THE POISSON EQUATION IN PARALLEL?

The computation of the right-hand side of the Poisson equation (3) can be done in parallel using the same constructs as used for (1)-(2) and (4). On a distributed memory system like the CM-5 the parallelization of the Poisson solver itself is more difficult.

The FFT/ICCG approach which works very well on the CRAY J932, performs poorly on the CM-5. The reason for this is twofold.

First of all, the code has to be modified for the CM-5. Indeed, the FFT part of the FFT/ICCG code on the J932 is parallel in the j-direction, and not parallel in the k-direction, while this is precisely reversed for the ICCG part. Consequently, on the CM5, this code will generate an enormous traffic of data between the parallel processors, either in the FFT part or in the ICCG part. The congestion caused by all this traffic can be avoided by redefining the parallel loops. On the CRAY the i-direction is used to vectorize. On the CM-5 vectorization is not so important as on the CRAY. All 64 vector units of the CM-5 work on fixed vectors of length 8. Since this length is relatively short, the i-loop can be used to parallelize the code on the CM-5. Then the FFT and the ICCG part have the same parallel directions, and an excessive traffic of data between the processors is avoided.

The second cause for the disappointing performance of the FFT/ICCG Poisson solver on the CM-5 is the speed at which ICCG iterations run. This point will be worked out in the remaining part of this section.

To start we consider the unpreconditioned Conjugate Gradient method for solving the pressure from the standard five-point discretisation on a uniform two dimensional grid. The data-parallel code for this reads

```
p = initial guess, r = initial residual,  s = 0,   beta = 0
rho = SUM(r*r)
WHILE ( rho .GT. tolerance ) DO
s = r + beta*s
q = - CSHIFT(s, DIM=1, SHIFT=-1) - CSHIFT(s, DIM=1, SHIFT=1)
    - CSHIFT(s, DIM=2, SHIFT=-1) - CSHIFT(s, DIM=2, SHIFT=1)
    + 4*s
alpha = rho/SUM(s*q)
p = p + alpha*s
r = r - alpha*q
rhon = SUM(r*r)
beta = rhon/rho
rho = rhon
END WHILE
```

Here, p, r, s and q are arrays (all have the same size as p has), and alpha, beta, rho and rhon are scalars. All variables are defined as double precision.

This code runs at about 25% of the peak of the CM-5. Its performance can be improved by a few percents by replacing the statement with the CSHIFT's by a call to a routine from the CMSSL library. With this Poisson solver the overall speed of a 2D DNS code lies a little over 500 Mflop/s on a 16 node CM-5. In three spatial dimensions, of which one is statistically homogeneous, a FFT/CG method runs also at such a speed.

The CG-algorithm has two synchronization points, namely the two innerproducts in the computation of alpha and beta. There are various approaches reported to reduce the costs of these two innerproducts. Demmel *et al.* (1993) propose to postpone the update of p one iteration. Then, this update can be overlapped with the computation of alpha, and thus the iteration has one synchronization point less. The resulting method has the same numerical stability as the standard CG. We have implemented the CG-method with postponed update of p on the CM-5, and found that it is not faster than the standard implementation: the data parallel compiler does not recognize the possibility to overlap one innerproduct with an array-update.

The shift to flop ratios for the CG-algorithm are 0.21 and 0.26 in two and three dimensions respectively. It is often remarked that the load plus store to flop ratio of CG is not very good. A closer look at the code generated by the data parallel CM-Fortran compiler shows that this ratio is not at all that bad: it is equal to 1.0 in 2D, and 0.9 in three dimensions.

As already remarked, the speed of convergence of the CG iteration can be improved by introducing an appropriate preconditioner. Here, an incomplete Choleski factorization is used as a preconditioner. This preconditioner introduces a recursion in both directions over the grid, but not along diagonals of the grid. Thus, in a diagonal-wise ordering the unknowns can be computed in parallel. This observation is explored on vector-computers: the vectors correspond to diagonals of the grid; see Section 5. On a parallel computer with local memory each processor can compute a part of a diagonal, if the unknowns are ordered diagonal-wise. Suppose that x is a square array of N^2 elements. To store the elements of x diagonal-wise we define an array xd of N*2N elements. The first diagonal (corresponding to d=1) of x is stored in the first column of xd, the second diagonal of x (d=2) is stored in the second column of xd, and so on.

When all arrays are stored in this manner, the data parallel code for the recursive relation reads

```
xd( ;1) = rd( ;1)
DO d = 2, 2*N
xd( ;d) = rd( ;d) - ad( ;d)* xd( ;d-1) -
        bd( ;d)*CSHIFT(xd(;d-1), SHIFT=-1))
ENDDO
```

The diagonal re-ordering has several drawbacks. Owing to the variations of the lengths of the diagonals, some processors perform superfluous computations. Obviously, this slows the computation down. This slowing-down is further strengthened by the fact that the vector units of the CM-5 always have to work on fixed chunks of 8 elements.

Yet, the most serious drawback of the above approach is formed by the communication costs. The data-flow for a diagonal-update is sketched in Figure 6.

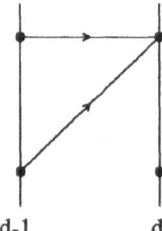

Figure 6.   The data-flow introduced by the incomplete Choleski preconditioner, when the unknowns are diagonalwise re-ordered.

To update the elements of the diagonal stored in the d-th column of xd, a CSHIFT of the previous diagonal d-1 has to be performed. The number of data-elements to be moved for this shift is relatively low. Consequently, the latency, i.e. the time for setting up the communication, dominates the communication time. In practice, the communication costs are excessively high: the CSHIFT in the DO-loop that computes xd(;d) takes almost all of the wall clock time. This DO-loop causes the preconditioned CG method to run at only a few percent of the peak of the CM-5. Consequently, this preconditioned CG method cannot compete with the unpreconditioned CG method. Moreover, in three spatial dimensions, of which one is statistically homogeneous, a FFT/ICCG method cannot compete with a FFT/CG.

It may be remarked that the long time required by the CSHIFT is partly due to a non-optimal implementation of this operation on the CM-5. Yet, even if the CSHIFT would be implemented optimally fast, the estimated time (based on the hardware of the CM-5) for communication is still too large for the diagonally re-ordered ICCG method to be competitive to the unpreconditioned one.

7. SOR strikes back

Having read the preceding sections one may wonder: Is there still a role for good-old SOR in this era of HPC, an era dominated by preconditioners, Krylov subspace methods and multi-grid? Therefore, in this final section we will discuss an application where we challenge these fancy methods to oust SOR on parallel computers.

7.1. PROBLEM DESCRIPTION

Our application concerns sloshing of incompressible liquid inside moving containers. Research started around 1980 when one of the authors was employed at the National Aerospace Laboratory NLR, where he became interested in analysing the influence of contained liquid motion on the dynamics of spacecraft. A combined experimental/theoretical research program was defined. On the theoretical side CFD techniques have been developed to predict the liquid motion (Veldman and Vogels, 1984). These techniques had to be capable of treating liquid regions with arbitrarily changing topology. Therefore a fixed Eulerian grid approach has been chosen, in which the liquid was allowed to move freely under the influence of inertial and capillary forces. For keeping track of the free liquid surface the VOF method was used (Hirt and Nichols, 1981).

The equations of motion are employed only within the liquid region. At the free surface boundary conditions are available for the pressure and the velocity (expressing continuity of the normal and tangential stress components). The boundary condition for the pressure is of Dirichlet type, and is implemented at the free-surface location through interpolation between two pressure nodes. Due to this condition the discrete Poisson matrix will no longer be symmetric; it can possess one or more complex eigenvalues with nonnegligible imaginary part. The shape of the liquid region can change arbitrarily, i.e. the coefficients of the Poisson matrix are not constant in time. In order to make at least the data structure invariant, in the pressure nodes not involved in the liquid dynamics equations are added of the form $p_{i,j} = 0.0$.

In the examples given earlier in this chapter the Poisson matrix remained constant throughout the full simulation. Thus the effort for making a preconditioner or a decomposition can be amortized over a large number of time steps. However, in the present sloshing appplication a preconditioner will have to be refreshed regularly, making this step more costly.

From the previous time steps an initial estimate can be made for the pressure: when the motion is very quiet, e.g. when the fluid is close to equilibrium, this estimate is very good. In these situations a simple method like SOR would require only a few iterations. When the motion is violent this initial estimate is less useful, and this is the situation where modern methods could show off.

It was decided to solve the Poisson equation with an SOR method with adaptive relaxation factor. If the free surface would not be present an optimal relaxation factor could be found theoretically. However due to the irregular equations near the surface (and the corresponding complex eigenvalues) the iterations with this factor will often diverge. Therefore the iterations with the optimum factor are monitored, and when signs of divergence appear a number of iterations with $\omega = 1$ are included to suppress the diverging components. Details of this method can be found in Botta and Ellenbroek (1985).

7.2. ALGORITHM AND IMPLEMENTATION

The basic structure of the algorithm extends that as described in the preceeding sections. A time step consists of three steps:

1. Compute an auxiliary velocity field (2);
2. Solve for the pressure (3) and find the velocity (4);
3. Determine the new location of the fluid.

Because of the changing liquid position the algorithm repeatedly has to check whether a cell is inside the liquid, near the liquid surface or away from the surface. In the routine that computes the auxiliary velocity this induces a small number of IF-statements. The routine that computes the coefficients for the Poisson matrix also contains some of these checks. But the routine that moves the free surface (VFCONV) is full of branching. In the early days of vectorcomputing, on the Cyber 205 e.g., this routine could not be vectorized by the compiler. Special machine instructions had to be written, which yielded an improvement of 3.5 on a small grid (8x16 grid points; much finer grids could not be handled in those days). For nostalgic reasons we will show a small part of the VFCONV routine in plain Fortran (as it is used now on the Cray J932, where it automatically vectorizes and parallelizes) and in Cyber 205 vector instructions. Figure 7 clearly shows that compilers have become much more intelligent since the early eighties.

The intricate complexity of the VFCONV routine seriously prevents approaching the maximum level of vectorization and parallelization. Hence, on a parallel machine the presence of this routine will influence the overall performance of the algorithm and it in particular influences the criticality of the Poisson solver. We will demonstrate this in the sequel.

The SOR Poisson solver has been formulated in a one-dimensional red-black ordering, such that on small grids, say 32x32, sufficient vectorlength is available for attaining full vector speed (which lies close to 150 Mflop/s at a theoretical peak of 200 Mflop/s). As a consequence there is only one single loop. Since the Cray J932 only parallelizes outer loops automatically, explicit compiler instructions now are necessary to enforce parallelization (Figure 8).

```
      FY1 = FN(I,JA)*ABVY + AMAX1((1.0-FN(I,JA))*ABVY-
     1                          (1.0-FN(I,JD))*DELY(JD),0.0)
      FLUXY = AMIN1(FY1,FN(I,JD)*DELY(JD))



      call q8vxtov(,,jd(1;ijmax),,fn(1;ijmax),,fnd(1;ijmax))
      call q8vxtov(,,ja(1;ijmax),,fn(1;ijmax),,fna(1;ijmax))
      call q8vxtov(,,jd(1;ijmax),,dely(1;ijmax),,delyd(1;ijmax))
      tmp(1;ijmax) = (1.0-fna(1;ijmax))*abvy(1;ijmax) -
     1                          (1.0-fnd(1;ijmax))*delyd(1;ijmax)
      cw(1;ijmax) = tmp(1;ijmax).gt.0.0
      call q8maskv(x^08^,,tmp(1;ijmax),,,cw(1;ijmax),fy1(1;ijmax))
      fy1(1;ijmax) = fna(1;ijmax)*abvy(1;ijmax)+fy1(1;ijmax)
      tmp(1;ijmax) = fnd(1;ijmax)*delyd(1;ijmax)
      cw(1;ijmax) = fy1(1;ijmax).lt.tmp(1;ijmax)
      call q8maskv(,,fy1(1;ijmax),,tmp(1;ijmax),cw(1;ijmax),fluxy(1;ijmax))
```

Figure 7. Two statements from the VFCONV routine in plain Fortran as used on the Cray J932, and their counterpart in Cyber 205 vector instructions. These statements take care that the amount of fluid to be fluxed through a cell face does not exceed the available free space in the acceptor cell (first statement), nor should it exceed the available amount of fluid in the donor cell (second statement).

```
          TDELR=0.0

CMIC$ PARALLEL
CMIC$1   SHARED(OMEG,NX2,NR,PR,DIVR,CWR,CER,CSR,CNR,TDELR)
CMIC$2   SHARED(NB,PB,DIVB,CWB,CEB,CSB,CNB,TDELB)
CMIC$3   PRIVATE(I,J,DELTAR,DELTAB,DIFF)

          DELTAR=0.0

CMIC$ DO PARALLEL NUMCHUNKS(32)
          DO 130 I=1,NR
             DIFF = - PR(I) + DIVR(I)
     1            - CWR(I)*PB(I-1)    - CER(I)*PB(I)
     2            - CSR(I)*PB(I-NX2-1) - CNR(I)*PB(I+NX2)
          PR(I) = PR(I) + OMEG*DIFF
          DELTAR = DELTAR + DIFF**2
 130   CONTINUE
CMIC$ GUARD
          TDELR = TDELR + DELTAR
CMIC$ END GUARD
CMIC$ END PARALLEL
```

Figure 8. Pressure update in the red cells with compiler directives that are necessary because the Cray does not automatically parallelize inner loops.

7.3. PERFORMANCE

We will now present some timing results of our sloshing code on various platforms:
– a HP 755 workstation with cache-based RISC architecture;
– a Cray J932 existing of 32 computing nodes with vector architecture.
Two examples will be discussed: in the first one a fine grid is chosen with 64x128 grid points for a realistic flow simulation which features a mix of strong and weak liquid motion. In the second example we largely exaggerate the influence of the Poisson equation by using a 1024x1024 grid (on which only a few time steps have been performed). In Tables 6 and 7 we give an impression of the relative contribution of the Poisson equation to the complete simulation.

TABLE 6. Relative processing time for SOR Poisson solver over a number of time steps on a 64x128 grid.

	HP-755	Cray J932 (clock cycles)			
	(secs.)	1	2	4	16
Poisson	286	22	11	6	2
Rest	122	21	16	14	12
Total	408	43	27	20	14
% Poisson	70%	51%	41%	26%	14%

The realistic example documented in Table 6 shows that on a RISC architecture typically 70% of the computational effort is spent on the Poisson equation. I.e. replacing it with an infinitely fast one a factor of three can be gained. On the vectorprocessor this percentage reduces to about 50%, allowing a gain of at most a factor of two. This decrease can be understood from the almost perfect vectorization of the Poisson equation, whereas other routines vectorize worse. Using parallel processing the contribution of the Poisson equation further reduces. On 4 processors the gain to be achieved by replacing SOR with an infinitely fast solver has become irrelevant already.

The second, exaggerated example (Table 7) shows a Poisson contribution of 98% on the RISC processor. Here a factor of 50 can be gained, which certainly makes it worthwhile to replace the SOR routine (one would be a fool not to do it with so many fancy methods around). Also, the factor of 10 to be gained on one vectorprocessor of the Cray cannot be ignored. But on 32 processors the routine VFCONV starts to dominate the program; the Poisson contribution has shrunk to 35%, and, even in this highly exaggerated example, there is nothing significant to gain anymore by replacing the SOR solver with a more modern one. Notice the perfect scaling of the SOR Poisson solver right up to 32 processors (the Cray analysis tools show a scaling of 31.1 out of 32).

TABLE 7. Relative processing time for SOR Poisson solver on an excessively fine 1024x1024 grid.

	HP-755 (secs.)	Cray J932 (clock cycles)					
		1	2	4	8	16	32
Poisson	493	79	40	20	10	5	2.6
Rest	9	8	6	6	6	5	4.8
Total	502	87	46	26	16	10	7.4
% Poisson	98%	91%	87%	77%	62%	51%	35%

This application shows that there is still a role to be played by SOR in the HPC era. We can only think of one argument to look for an alternative: When SOR is using all 32 processors, noone else can use the machine. If there would be an algorithm requiring comparable wall-clock time using less processors, then, while we are doing our computations, others could use the computer too.

Acknowledgements

Both the Stichting Nationale Computerfaciliteiten (National Computing Facilities Foundation, NCF) with financial support from the Nederlandse Organisatie voor Wetenschappelijk Onderzoek (Netherlands Organization for Scientific Research, NWO) and the National Aerospace Laboratory NLR are gratefully acknowledged for the use of supercomputer facilities.

References

Botta, E.F.F. and Ellenbroek, M.H.M. (1985) A modified SOR method for the Poisson equation in unsteady free-surface flow calculations. *J. Comp. Phys.* 60, pp. 119–134.
Demmel, J.W., Heath, M.T. and Van der Vorst, H.A. (1993) Parallel numerical linear algebra. *Acta Numerica* 2, pp. 111–199.
Dongarra, J.J., Duff, I.S., Sorensen, D.C. and Van der Vorst, H.A. (1991) *Linear System Solving on Vector and Shared Memory Computers*. SIAM, Philadelphia.
Eisenstat, S. (1981) Efficient implementation of a class of preconditioned Conjugate Gradient methods. *SIAM J. Sci. Stat. Comput.* 2, pp. 1–4.
Gustafson, I. (1978) A class of first-order factorization methods. *BIT* 18, pp. 142–156.
Harlow, F.H. and Welsh, J.E. (1965) Numerical calculation of time-dependent viscous incompressible flow with free surface. *Phys. Fluids* 8, pp. 2182–2189.

Hirt, C.W. and Nichols, B.D. (1981) Volume-of-Fluid (VOF) method for the dynamics of free boundaries. *J. Comp. Phys.* **39**, pp. 201–255.

Kim, J., Moin, P. and Moser, R.D. (1987) Turbulence statistics in fully-developed channel flow at low Reynolds number. *J. Fluid Mech* **177**, pp. 133-166.

Meijerink, J.A. and Van der Vorst, H.A. (1977) An iterative solution method for linear systems of which the coefficient matrix is a symmetric M-matrix. *Math. Comp* **31**, pp. 148–162.

Moser, R.D. and Rogers, M. (1993) The three-dimensional evolution of a plane mixing layer: pairing and transition to turbulence. *J. Fluid Mech.* **247**, pp. 275-320.

Reynolds, W.C. (1990) The potential and limitations of direct and large eddy simulations. In: *Whither Turbulence? Turbulence at the Crossroads.* J.L. Lumley (ed.). Springer-Verlag, Berlin, pp. 313–343.

Spalart, P.R. (1988) Direct simulation of a turbulent boundary layer up to $R_\Theta = 1410$. *J. Fluid Mech.* **187**, pp. 61-98.

Van Loan, C. (1992) *Computational Frameworks for the Fast Fourier Transform* Frontiers in Applied Mathematics: 10, SIAM, Philadelphia.

Veldman, A.E.P. and Vogels, M.E.S. (1984) Axisymmetric liquid sloshing under low gravity conditions. *Acta Astronautica* **11**, pp. 641–649.

Verstappen, R.W.C.P. and Veldman, A.E.P. (1994) Direct numerical simulation of a 3D turbulent flow in a driven cavity. In: *Computational Fluid Dynamics '94*, Wagner *et al.* (eds.). John Wiley and Sons, Chichester, pp. 558–565.

Verstappen, R.W.C.P. and Veldman, A.E.P. (1996) A fourth-order finite volume method for direct numerical simulation of turbulence at higher Reynolds numbers; to appear in: Proceedings of ECCOMAS '96.

HIGH PERFORMANCE COMPUTING: TRENDS AND EXPECTATIONS

P.H. MICHIELSE
Hewlett Packard Company
Convex Computer BV
Europalaan 514
3526 KS Utrecht
The Netherlands

Abstract

In this paper we try to identify trends and to forecast expectations in the field of High Performance Computing (HPC). Since trends can only be recognized by observing historical data, this paper contains quite some information on past and current developments in HPC. Using these data, we try to give some insight in to-be-expected hardware and software. All aspects that together constitute HPC will be discussed: processors, memory, architectures, parallelism, message-passing, cache coherency, compilers, etc. Also developments in the area of Computational Fluid Dynamics will be touched upon, in relation to HPC.

1. Introduction

Historically, one could say that the field of High Performance Computing (HPC) started in 1976, when Cray Research Inc. introduced the Cray-1 vector supercomputer. The word "supercomputer" was used to characterize this machine, which was based on high-speed vector technology and large and fast DRAM memory. Since then, the field of HPC expanded through the introduction of multi-processor vector supercomputers, massively parallel processors, clusters of RISC workstations and, as the latest category, scalable parallel processors. The fact that, apart from expansion, HPC has continually been undergoing change, is probably best illustrated by the number of hardware vendors that started in, and disappeared from HPC. Among them are some well known names as Alliant, Cray Computer Corporation, Kendall Square Research Inc. and Thinking Machines Corporation. Although in recent years the rate of change of HPC has increased dramatically, it seems that there is some convergence. In this paper, we will try to address the direction of convergence by investigating trends and expectations for the future in HPC. This will be done by discussing the individual components that, together, constitute HPC systems: hardware (processors, memory), architecture (parallelism), software (programming models, compilers) and algorithms. In particular, algorithms and implementations related to Computational Fluid Dynamics (CFD) will be considered.

261

P. Wesseling (ed.), High Performance Computing in Fluid Dynamics, 261–278.
© *1996 Kluwer Academic Publishers.*

2. Hardware

2.1. VECTOR PROCESSORS

It is no surprise to state that the speed of the processors used in HPC is increasing. The aforementioned Cray-1 supercomputer's cycle time was 12.5 nanoseconds (ns), in 1976. The cycle time of its successors and of competitive vector machines have gradually decreased to roughly 2 ns in 1996 (cf. Figure 1). This holds for the high-end systems of all four traditional vector computer vendors: Cray, Fujitsu, Hitachi and NEC. The factor of 6 to 7 increase in vector processor speed in 20 years is not really dramatic. It is expected that only modest gains in processor cycle time can be achieved against very high costs. The speed of light, coupled with the distance that needs to be covered by the signals, limits the size of the processor, leading to cooling problems. These physics-related limits, and the associated high costs to approach the limits as close as possible, will limit the single processor cycle time to something close to 1 ns.

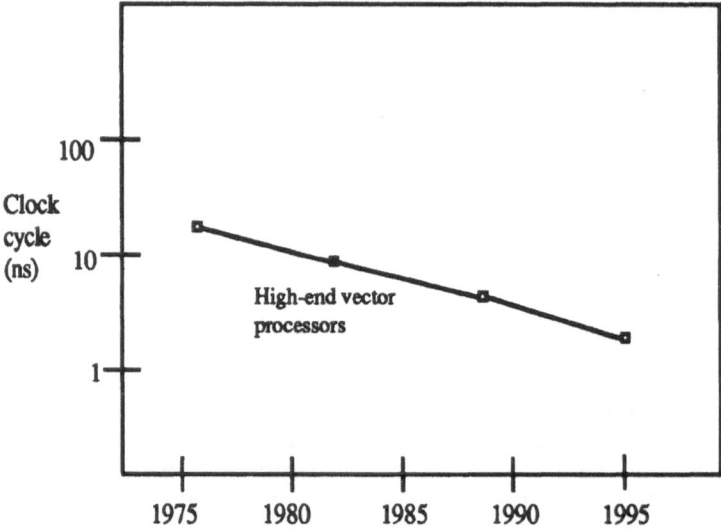

Figure 1.   Historical behavior of vector processor cycle time.

When the Cray-1 was introduced, its peak performance was 167 Mflop/s. This number could be achieved by the processor's ability to execute two floating-point calculations per 12.5 ns cycle. This concept became known as chaining: the result of one calculation (e.g. multiplication) was immediately transferred to another so-called functional unit (e.g. addition), where it could be processed further. The segmented functional units support pipelining, so that effectively two floating-point results per cycle could be delivered. In order to improve performance without facing the need to reduce processor cycle times, the concept of multiple pipes became popular. Instead of one set of functional units per vector processor, each vector processor was equipped with two, or even four, sets of functional units. This introduced fine-grained parallelism within the processor. For more details on vector processing, we refer to [1] and [2].

In the 1980's, users became more and more familiar with vector processors, and vector computing found its way into mainstream (scientific) computing. The market acceptance of supercomputing paved the way for the success of mini supercomputer vendors like Alliant and Convex, who developed and manufactured much cheaper (but slower) vector processors. Also in this period, multi-processor supercomputers were developed. A great deal of scientific software was ported to, and optimized for, vector processors, contributing to the success of both the big supercomputers and the minisupers.

However, problems developed in the early 1990's. Very high development costs for classic supercomputers and the increasing performance of single-chip CPUs made the price/performance ratios of traditional vector computers not compare well against these single-chip computers. Now, in 1996, this is still the case and it is not expected that (multi-processor) vector computers will be the computer architecture of the future. However, due to their specific capabilities, vector computers will still be very useful for a certain, probably limited, set of applications.

2.2. RISC PROCESSORS

As was remarked at the end of the previous section, single-chip CPUs have become more and more competitive to vector processors. Figure 2 shows the increase in speed of single-chip CPUs, compared to vector processors.

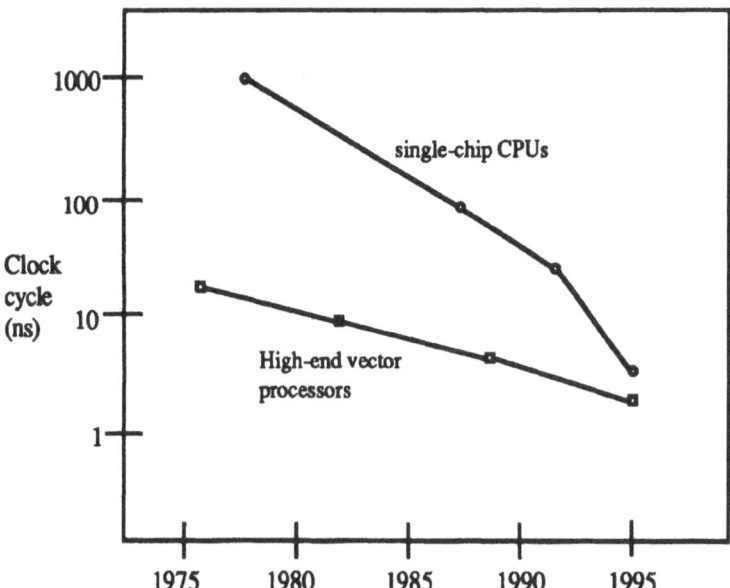

Figure 2. Historical behavior of vector processor and single-chip CPU cycle time.

As is clear from Figure 2, single-chip CPUs have experienced dramatic progress in clock cycle in 15 years: a factor greater than 100. The speed of single-chip CPUs have increased that much through a combination of three aspects: higher levels of integration on the chip (more components on a single die), reduced design rules (basically the width of the conductor paths on the chip) and the concept of RISC: Reduced Instruction Set Computing. The idea of RISC was to drastically lower the number of clock cycles needed per instruction. This could be achieved by providing fewer and simpler instructions. More complex instructions consisted of a few primitives. Also due to technological improvements, RISC processors became very popular, and, as Figure 2 shows, approach the cycle time of traditional vector processors. RISC processors are currently used in desktop computing (workstations), which gives the volume to cover the huge development costs of new (faster) processors. Today it costs approximately one billion US dollars to construct a facility to produce fine design rule chips. Examples are Digital (Alpha, $700 million), Intel (Pentium, $1 billion and P6, $1.2 billion) and Intel/HP (P7 or PA9000, $1.5 billion projected). In order to cover these huge costs, large companies are partnering to develop next-generation microprocessors. Examples are: IBM/Motorola/Apple and Intel/HP.

However, it is not expected that the clock cycle of new RISC processors will be reduced with more than a factor of 10 again. Currently, Digital pushes the clock cycle the most (> 300 MHz, > 600 Mflop/s). HP (> 180 MHz, > 720 Mflop/s) and MIPS (> 200 MHz, > 400 Mflop/s) have just released their latest RISC chips. IBM lags a little behind at the moment. As was the case with vector processors, clock cycle limits are forecasted at somewhere around 1 ns.

Similar to vector processors, which use multiple pipes, modern RISC processors employ multiple functional units as well. Very common are separate functional units for addition and multiplication. Currently, the maximum number of (floating-point) functional units on a chip is 4, which might be extended to 8 in the future. Typically, codes developed for vector computers could benefit from this. On the other hand, it is more difficult to obtain a high percentage of peak performance with more functional units: there needs to be enough (simultaneous) floating-point work to feed the functional units. Figure 3 shows a high-level block diagram of a RISC processor.

Probably more important than feeding multiple floating-point functional units is the ability of RISC processors to perform multiple load/store operations into registers. So far, there are not many RISC processors that implement this strategy. It is to be expected that in the near future this aspect will get more and more attention. The current drawback of computers based on RISC processors is their poor memory bandwidth, i.e. it takes a relatively long time to load data from memory (via caches) into the processor. In subsequent sections of this paper, this aspect will be discussed in detail.

Another aspect of RISC processors is the so-called "latency in the processor". Suppose the result of an addition or multiplication is written into a register. On most RISC processors, this result cannot be used in the next clock cycle. In general, it takes a couple of cycles (depending on the chip and the particular instruction) before the result can be used in some other computation. This is called "the latency in the processor". Some latencies (e.g. division and square root) can be really large: tens of clock cycles.

New features which are expected to show up in the next generation RISC chips (and which are already there in some of them) include out-of-order execution and speculative instruction execution. Out-of-order execution means that the hardware tries to re-order instructions for efficient (multiple) functional unit usage. Special Instruction Reorder

Buffers are in the hardware to enable the processor to (re-)schedule instructions as efficiently as possible. Out-of-order execution is closely related to Very Large Instruction Word (VLIW) processing, which will be touched upon in the section on compiler developments. VLIW processing is a typical example of compiler developments and features that move from the compiler to the hardware. Note that this has always been the case, especially in vector processing. In this sense, there is nothing really new. In speculative instruction execution, the hardware tries to predict the instruction flow, yet another example of a compiler feature that transfers to the hardware.

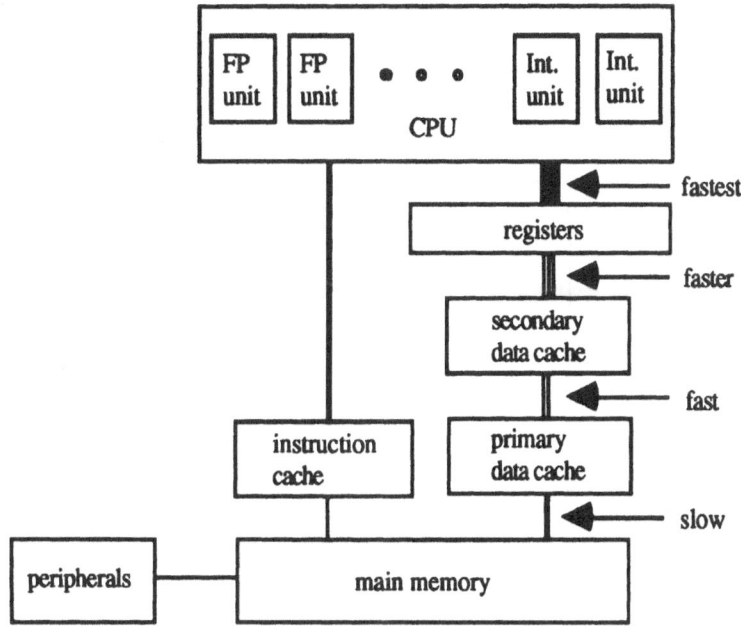

Figure 3.   High-level block diagram of a RISC processor.

In this section we have tried to outline the state-of-the-art and to-be-expected characteristics of RISC processors. We have observed the following trends:

- Clock cycles continue to decrease, (semi-conductor) limit around 1 ns;
- Multiple functional units, maximum 4 to 8;
- Superscalar execution, i.e. the ability to execute two or more instructions simultaneously;
- Need for addressing latencies in the processor;
- Former compiler techniques (VLIW, branch prediction) move to the hardware.

The bottom line is that RISC chips are becoming more and more complex, which is reflected by the increasing manufacturing costs. Only volume production can justify these costs, which is the (economic) reason that RISC chips show up in (massively)

parallel computers. Since the performance of single processors is limited (although the
limits have not been reached yet), the only way to substantial performance gains is
through (massively) parallelism. Before turning to parallel architectures in detail, we
first discuss another important hardware aspect: memory.

2.3. MEMORY

In the previous subsection we observed that RISC processor speeds have made
substantial progress. Unfortunately, the same is not true for memory. The memory of
most computers is comprised of DRAM chips. Two aspects of memory are important
when considering the progress made in integrated circuit technology: the density (bits
per chip) of DRAMs and the speed of DRAMs. Figure 4 shows amazing progress with
respect to density: a factor of 1000 increase in the density in 15 years. Currently, 64
Mbit DRAM chips are (nearly) here, and it is expected that densities will grow with
about the same rate into the next century. However, the DRAM access time (the
"speed") is of more concern to the user. Figure 5 shows the decrease in access time for
DRAMs, for the same period as the density. Here we observe a factor of about 5 only.

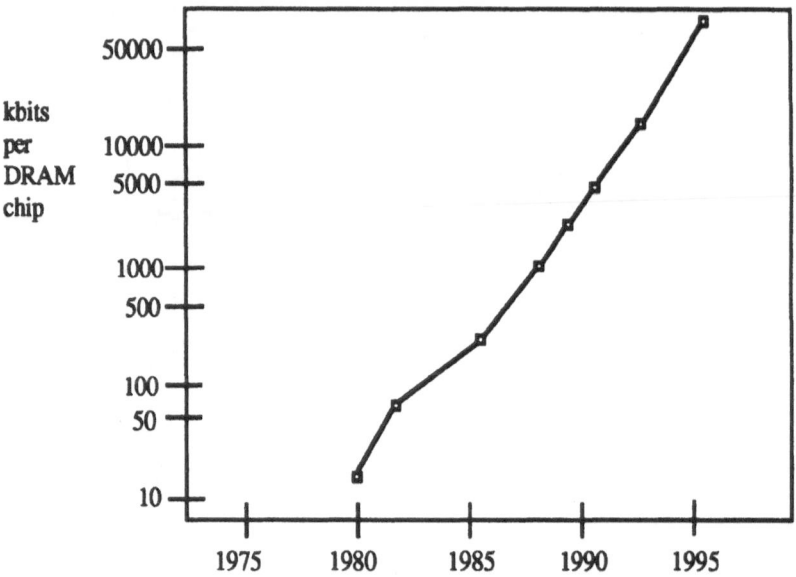

Figure 4.   Historical behavior of DRAM density.

It is clear from Figure 5 that DRAM access time is not keeping pace with advances
in processor frequency. Since it is not expected that this will change dramatically, the
"distance" (here distance is taken to mean the number of processor clock cycles) to
memory is getting larger and larger.

The memory latency, as discussed above, is an important issue when discussing
"distance" to memory. The other related characteristic is the sustainable memory
bandwidth. For the classic supercomputer, a significant portion of the cost is in the
memory subsystem. RISC processors, originally designed for the high-volume desk-

side market, spend a smaller amount and percentage on the memory subsystem. The memory bandwidth for high-end vector computers is on the order of several GByte/s, while for RISC-based workstations it is in the order of several hundreds of MByte/s, a difference of an order of magnitude.

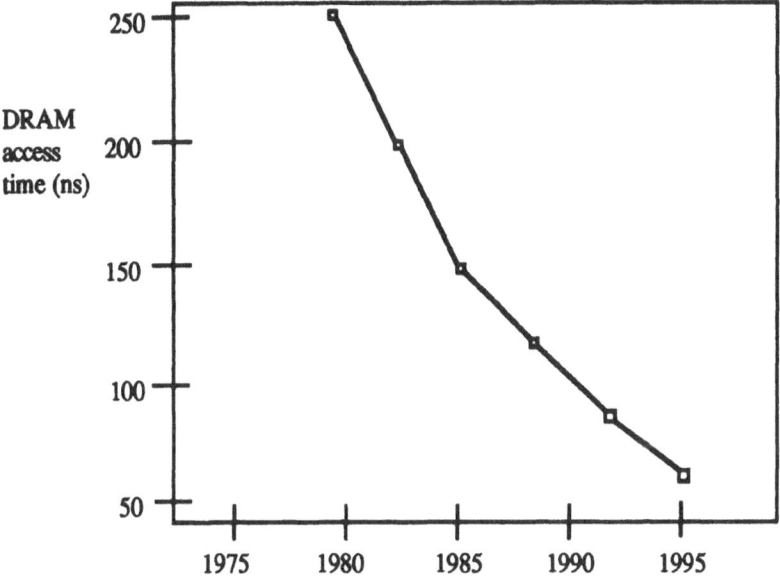

Figure 5.   Historical behavior of DRAM access time.

In order to decrease the striking difference between the memory subsystems of vector computers and RISC-machines, RISC vendors have invented caches: "intermediate" memory between CPU and main memory. The idea is to provide the processor with memory that is much closer than main memory. When the processor needs data, it copies the data from main memory into the cache. This operation suffers from the poor memory bandwidth, but after that the processor operates on that particular data in the cache. As long as the processor can operate on data that is in the cache, memory latencies and bandwidth are more favorable, and performance is good. However, as soon as the required data is not in the cache, the processor has to load the data from main memory, leading to (considerable) performance degradation. It follows that knowledge on the way the cache operates is crucial for performance. Two kinds of cache strategies are used: direct-mapped caches and n-way associative caches.

In direct-mapped caches, the mapping between the location of the data in main memory and in the cache is fixed: the (either virtual or physical) address of the data determines the position of the data in the cache. This has the consequence that data with different addresses may map to exactly the same cache position, without any flexibility. This is not the case in n-way associative caches: each data element in main memory can be mapped to any of n positions in the cache. A typical value of n is 4, although there are also caches in which data from main memory can be mapped to any location in the cache: fully associative cache.

Currently, it is hard to say which cache strategy will be the winning one. Direct-mapped caches are much simpler and, for a given size, faster than n-way associative caches. Other factors that determine the type of cache to be used include the speed and the density of the SRAMs used to construct the cache. There does not seem to be convergence among the RISC platforms into one of the strategies. Both direct-mapped and n-way associative caches are being used. Cache sizes currently vary from 128 kBytes to several MBytes.

What seems to be of more importance are the latency and bandwidth between main memory and cache. At the moment, there are factors of about 10 to 30 between loading data out of cache and loading data out of main memory. State-of-the-art RISC chips, like the MIPS R10000 and the HP PA8000, try to hide much of the latency and have improved bandwidths. One strategy is to support a number of simultaneous, independent loads from memory into the cache. This can be done by issuing loads from memory into the cache ahead of time, i.e. before the data is actually needed. Another strategy that is used is the following. When a cache miss is taken, not only the relevant cache line is being loaded out of main memory, but also the consecutive cache line. This is based on the expectation that the data in a consecutive cache line will be needed by the processor very soon as well. A cache line is the unit of transfer between main memory and cache, and is typically something between 32 and 128 bytes. Note that these loads proceed without necessarily stalling the processor.

This strategy of having multiple simultaneous loads (on the HP PA8000 a maximum of 10 cache lines, 32 bytes each, and on the MIPS R10000 a maximum of 4 cache lines, 128 bytes each) will especially boost the performance of codes that heavily use large array operations, i.e. vector codes. In a certain sense, the (future) memory subsystem of RISC machines can be characterized as converging towards the memory subsystem of classical vector computers, with the additional advantage that the RISC processor itself does not distinguish between vectorizable code or not, while a vector processor does.

What has not been discussed so far is the trend towards 64-bit implementations of RISC processors. The advantage of 64-bit processors lies in the expansion of the address limitation inherent in 32-bit chips. Theoretically, 32 bits can address 4 GBytes of (virtual) memory. In practice, user programs are able to address 2 GBytes of (single-processor) memory. The 2 GByte address limit of 32-bit architectures is starting to be exceeded by some applications. For this reason, 64-bit implementations of RISC chips are being developed now. Demand for 64-bit chips comes from the commercial market as well. Especially large database applications will certainly realise considerable performance improvement when being able to address more than 4 GBytes. At the moment, Digital's Alpha processors and SGI's MIPS processors are 64-bit implementations (also Cray's vector processors). The other vendors will follow (or have already followed) with 64-bit implementations of their processors as well. Except for Digital with Alpha (and Cray), all incorporate a 32-bit mode to permit existing 32-bit applications to be readily executed. Although it will take time to port applications to full 64-bit mode, 64-bit addressing is definitely needed in HPC, since it removes the 2 GByte address limit.

In this subsection we have surveyed trends in computer memory developments. Some trends are:

- Increase of (currently poor) memory-to-processor bandwidths;
- Density of DRAM chips continues to grow;
- Demand for large (virtual) memory address space supported by 64-bit addressing.

3. Parallel Architectures

3.1. MIMD SYSTEMS

In the previous section we have investigated hardware trends. The increase in speed of (RISC) processors based on semi-conductor technology is limited, and for that reason the general expectation is that only parallelism will be able to boost performance significantly. Flynn [3] has subdivided parallel architectures into several classes, and it is generally recognized that so-called MIMD (Multiple Instruction Multiple Data) machines have the most general-purpose potential. Up to a few years ago, there were two kind of MIMD architectures: shared memory (SM), with parallel vectorcomputers as typical representatives, and distributed memory (DM), with clusters of workstations as typical representatives.

DM machines in general have the following characteristics:

- The building block is a processor with local memory;
- Potential for many processors;
- Easily expandable with respect to number of processors and size of memory;
- No parallel compiler technology;
- Message-passing programming paradigm;
- Interconnection network;
- Memory latency within a broad range;
- Each building block has its own operating system.

The memory latency on DM computers is closely related to the interconnection network. When a processor needs to access data that is not located in its own local memory, then the data must be moved through the interconnection network to the requesting processor by message-passing. Hence, to a large extent, the interconnection network may determine the overall application performance. Figure 6 gives a historical overview of the current interconnect bandwidths [5]. In this context, latency is the time needed to access or move zero bytes of data, i.e. it is the overhead.

As shown in Figure 6, the interconnect bandwidths between computers have significantly increased. This supports the idea that network computing (NOW: Network Of Workstations) may even provide supercomputer-level computational power. Indeed, under the right circumstances, the network-based approach can be effective in coupling several similar multiprocessors, resulting in a very powerful configuration. However, while bandwidths have increased significantly, this is not the case for latencies. As with DRAM access time, latencies have not been improved by the same factor as bandwidths. High latencies are the current drawback of large clusters of workstations from a performance point of view. However, there certainly are applications that run very well on clusters of workstations. Although DM machines are not known for their ease-of-use, there are some (more or less standardized) tools available, to be discussed in section 4. For this reason, we expect that DM computing in clusters of workstations will turn out to be a surviving architecture in the next few years.

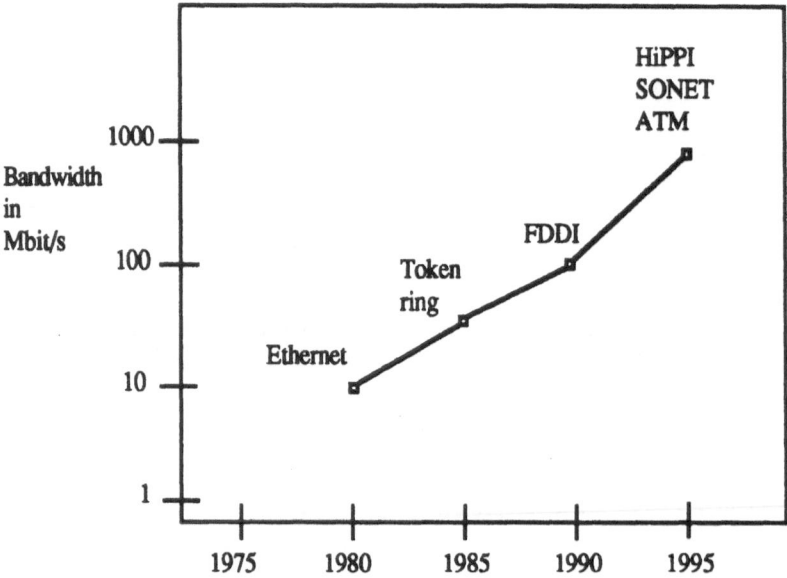

Figure 6.   Interconnect bandwidths.

Traditional SM systems have the following characteristics:

- Each processor can access the complete memory;
- More difficult to expand with respect to processors and memory;
- Parallel compiler technology;
- Shared memory and message-passing programming paradigms;
- Crossbar or bus technology;
- Uniform memory latency;
- One operating system for the entire machine;
- Easier to use than DM MIMD machines.

The general view is that SM computers are easier to use than DM machines. On the other hand, until recently, DM systems could be expanded with many more processors than SM machines.

A few years ago, some hardware vendors decided to combine the advantages of both SM and DM MIMD parallel computers. This has given rise to parallel computers with the following characteristics:

- All memory is globally addressable by any processor;
- Memory is physically distributed within the machine;
- Non-uniform memory latency;
- The building block is a symmetric multiprocessor (SMP).

Names such as Distributed Shared Memory (DSM) or Globally Shared Memory (GSM) are used to describe this type of machines. Figure 7 shows a typical DSM architecture.

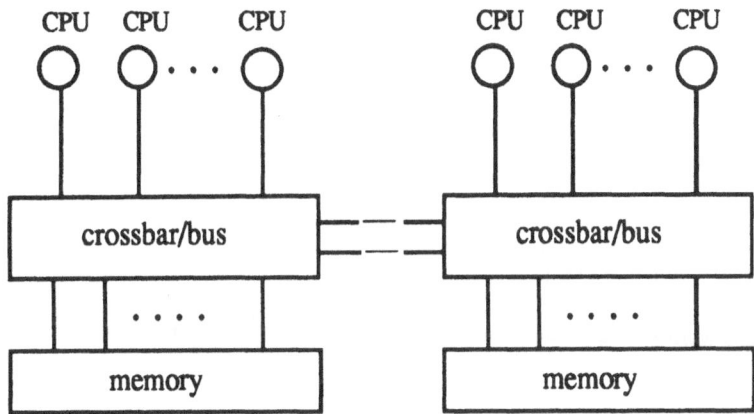

Figure 7. DSM architecture.

The building block (SMP) of DSM machines does not necessarily consist of more than one processor. One-processor building blocks exist, e.g. in the Kendall Square Research KSR1. Other examples are the Cray T3D (building block has 2 processors) and the HP/Convex Exemplar (building block has 8 processors). The most important characteristic of DSM machines is that each processor is able to access any memory location, without assistance of other processors (which is not the case in true DM architectures: message passing does require other processor's assistance). As we remarked earlier, this means that the memory latency is not uniform anymore: the DSM machine has a characteristic called Non Uniform Memory Access (NUMA). However, the range of memory latency has become much more narrow now, due to the extremely fast interconnect between the building blocks. Typically, the difference in latency between a processor accessing memory in its own building block and a processor accessing memory in another building block is less than a factor of 5. Both the HP/Convex Exemplar and the Cray T3E make use of SCI-interconnects, to be discussed later, between the building blocks.

An important numerical aspect with respect to implementation on DM or DSM machines can be illustrated by the following example. Consider the Poisson equation in two dimensions on the unit square, discretized by finite differences and solved by an iterative method like Gauss-Seidel or SOR. We apply domain decomposition to solve this problem in parallel. Implementation on DM machines (message passing) is in general such that each subdomain has overlap rows and columns with its neighboring subdomains. These overlap rows and columns are used to update values of the rows and columns at the boundaries of the subdomains. For the next iteration, these overlap rows and columns are exchanged between neighboring subdomains. In fact, this means that the Gauss-Seidel or SOR process most likely do not use the most up-to-date values in the overlap areas. This has consequences for the number of iterations needed to obtain numerical convergence.

On the other hand, on DSM systems there is no need for overlap areas, since each processor can access the complete computational grid. Hence, each processor is potentially able to use the most up-to-date values to update the numerical solution in its own grid points. Note, however, that the exact rate of numerical convergence depends on the actual update order within the subdomains. Experiments have shown that the number of iterations needed for numerical convergence on DSM is much lower than on DM machines. Even divergence was observed on DM machines, for only a small number of processors. This shows that from a numerical point of view, DSM systems have an advantage over true DM architectures.

It is expected that DSM architectures will find their place among the architectures of the future. Clear advantages are the fact that the machine is much easier to use, that "old" source code might transparently benefit from parallel compiler technology, and that the architecture is scalable. Scalability of the architecture can be realised through a distributed crossbar and through fast (SCI) links between the building blocks. Networks of workstations (DM computing) will certainly be useful for certain types of applications (embarrassingly parallel), but are somewhat less likely to become main-stream computing. However, due to their current market position and relatively easily expandable architecture, they will be around. Also, true SM machines, like parallel vector computers, will remain very useful for certain types of applications, but are not expected to be the architecture that will boost application performance by large factors.

3.2. CACHE COHERENCE

An important aspect of DSM parallel computers that has not been considered yet can be expressed by the following question: Does a parallel (DSM) application always see the entire machine or not? This question can be clarified with the following parallel execution example:

- CPU 1 modifies (shared) variable x
- CPU 2 reads   (shared) variable x

The question is: Does CPU 2 automatically see the updated variable x? If the answer to this question is "yes", then the DSM parallel computer is cache coherent. If the answer is "no", then the machine is not cache coherent. DM machines are not cache coherent.

The presence or absence of cache coherency in DSM computers is an important differentiator. In a non-cache-coherent DSM machine, the user needs to explicitly take care of coherency in his code (if he wants correct answers). This means that when some processor has updated shared data that might be needed by another processor, this data has to be forced from the data cache to shared memory. This has two disadvantages: user intervention is needed and it might reduce performance since data caches have to be flushed. In a cache coherent machine, the hardware needs to take care of each processor always seeing the most up-to-date data. Therefore, cache coherent machines offer easy programmability and a transparent way of (trying to) parallelizing millions of lines of legacy code. In our opinion, these reasons give cache coherent parallel systems better opportunities than non cache coherent parallel systems in the HPC arena.

For more details on cache coherency we refer to [4].

The last aspect of parallel architectures we would like to discuss is the combination of scalability and cache coherence. We have just illustrated that both for reasons of programmability and numerical behavior, cache coherence is important. However, what consequences has the feature of cache coherence on scalability of the underlying parallel architecture? Within SMP's, cache coherence is implemented in hardware. The actual implementation is such that due to the relatively small number of processors, SMP's are considered to be a scalable architecture. As DSM parallel machines may consist of a collection of SMP's, cache coherence within a node (i.e. within an SMP) does not need to be an issue. The point of concern of course is whether it is possible to maintain cache coherence between SMP's, without sacrificing performance of parallel programs when using more processors. Typically, this is what we consider as a scalable parallel architecture. It has turned out to be possible to maintain cache coherence and scalability in DSM parallel architectures. For instance, in HP-Convex Exemplar systems, a hardware-implemented directory structure is used to keep track of shared data that is used by processors throughout the machine. In this implementation, the interconnect between the nodes is crucial. The IEEE SCI (Scalable Coherent Interface) standard ([5]) has become a usable technology to implement cache coherence in a scalable way when many processors are used.

In our opinion, scalable cache coherency gives DSM machines the potential to have as many processors as true DM machines. The advantages of DSM architectures, like programmability and ease-of-use, over DM architectures will cause DSM to be the dominating architecture in the (less) near future.

4. Parallel Programming Models

4.1. MESSAGE-PASSING PROGRAMMING

In the previous section we have discussed parallel architectures to some extent. Distributed Memory (DM), Shared Memory (SM) and Distributed Shared Memory (DSM) architectures have been reviewed. Historically speaking, parallel systems with a large number of processors have always been DM machines. Only in the last few years, the concept of DSM has been developed and implemented.

On true DM machines, the parallel programming model is based on explicit message-passing programming. The first DM machines, like those from nCube, Intel and Meiko, implemented proprietary message-passing libraries, basically specifying the sending and receiving of messages between processes. These processes were mapped onto the processors, and together they constituted the parallel application. Although the concepts of these proprietary libraries in general were equal, this situation (no portability between platforms) was not considered as ideal. For that reason, attempts to standardize message-passing libraries (i.e. the programming interfaces) were initiated, with the ultimate goal to make codes that rely on message-passing more portable.

To this purpose, several "languages" have been developed. The most successfull language (or environment) at this time seems to be PVM: Parallel Virtual Machine [6]. PVM was developed at Oak Ridge National Laboratories and the University of Tennessee, PVM offers the basics of message-passing (sending and receiving of messages), together with some advanced features like process groups, the possibility of adding and removing processors to the virtual machine, etc. Being developed as a successor to PVM is MPI: Message Passing Interface [7]. MPI can be viewed as an improved and extended version of PVM. For instance, MPI does not need a daemon process anymore, as PVM does.

Currently, PVM and MPI have been adopted by all of the parallel computer vendors. Of course, true DM architectures, like the IBM SP-1 and SP-2 and other clusters of workstations, have implemented message-passing, since it is their only way to support parallelism. However, also vendors of traditional SM (and DSM) parallel systems, like Cray, HP/Convex and SGI, have accepted the importance of PVM/MPI based implementations, simply because there are so many PVM/MPI based applications. For this reason, full PVM/MPI implementations are available on SM and DSM computers as well.

On DM parallel machines, socket connections between processes are used to implement sending and receiving messages in a PVM or MPI environment. On SM or DSM machines, usage of sockets is possible as well, although vendors of these types of machines have found ways to use shared memory to implement message-passing. Since each processor of the DSM machine can access the entire memory, it is convenient to define certain regions in memory, which can be used by the parallel processes of a PVM/MPI application. For instance, when process 0 sends a message to process 1, it can physically copy its send buffer (which contains the data) into this region, after which it can be read by process 1. In this way, socket connections and sending/receiving messages are being replaced by memory copies. The advantage of such an implementation is that latencies are much lower and bandwidths much higher, as is shown in [8]. Hence, message-passing programming models, being essential on true DM parallel machines, can also be used, even more efficiently, on SM or DSM systems.

4.2. COMPILER DEVELOPMENTS

With PVM and MPI becoming de-facto standards for message-passing, serious attempts have been made to create portable message-passing libraries. In a similar spirit to message-passing, High Performance Fortran (HPF), [9], tries to offer the possibility of a truly portable programming language for parallel machines. This portability is intended across different machines and perhaps machine types. HPF is meant as an extension to Fortran, supporting data parallel programming, and aimed at top performance on diverse (MIMD or SIMD) parallel architectures. In order to accomplish this, HPF basically has to solve two problems: how to map computation to processors and how to formulate the communication implied in doing so. Through the use of compiler directives, the user should be able to specify how data has to be mapped on the machine. The HPF compiler then inserts system-specific message-passing calls, or probably memory copies on SM or DSM machines. The key to success of HPF is the availability of true HPF compilers, which does not seem to be going as quickly as anticipated at the initiation of the project. For that reason, it probably becomes more and more difficult for HPF to compete with true message-passing (PVM/MPI) applications and with traditional Fortran compilers. Another issue in this might be the fact that HPF currently seems to be designed for "regular" problems (regular grids), and is hardly suitable for irregular problems.

Historically, the introduction of the first Cray vector supercomputer needed to be followed by a compiler that could take advantage of the hardware features of vector processors. As a result, automatically vectorizing compilers have enjoyed a development period of nearly 20 years. One of the major concepts in automatically vectorizing compiler technology is the so-called strip mining principle: a vectorizable loop is split into chunks as large as the length of the vector registers of the underlying vector processor. Strip mining also paved the way for parallel vector processors: the chunks of the vectorizable loops could also be handled in parallel on the available vector processors of the computer. The hardware of parallel vector computers (like memory access, crossbar technology, etc.) is such that this so-called fine-grained parallelism is efficiently supported (i.e. gives speed-ups). So, automatically vectorizing compilers developed into automatically parallelizing compilers, adding more and more features to discover and implement coarser-grained parallelizations: parallel outer loops, compiler directives to execute subroutines in parallel, etc. Apart from these parallelization efforts, it can be noted that compiler features have gradually moved to the hardware. For instance, certain mathematical operations, like divide and square root, were originally executed in software, but have become hardware-implemented now.

The observation of compiler features moving to the hardware is also typical for current RISC machines. This also gives rise to the following question: where should the (optimization/parallelization) work be done: in hardware or in software (by the compiler)? We will try to elaborate on this question by considering the following example.

Consider a simple Fortran (or C) DO-loop. Compilers for RISC processors typically generate a branch instruction to compare the iteration count of the loop with the loop length. If the loop length is not yet reached, the next iteration of the loop is carried out, otherwise, the loop finishes and the next instruction needs to be executed. When mapping these instructions to the hardware of RISC processors, a problem arises. Since the processor does not know whether the branch is taken or not, the pipeline of instructions for the CPU is disturbed. The hardware approach is to either stall the CPU, evaluate the branch, and resume the pipeline, or to predict (statically or dynamically) the branch (and clean up if the prediction was wrong). The software solution to this problem is the "branch-delay slot": the instruction after the branch is always executed, regardless of the branch result. This requires that the compiler schedules the code correctly, otherwise wrong answers will result.

This example is typical for the current situation in HPC: performance improvements can be reached by both compiler developments and architectural improvements. In some sense, they are even complimentary. Another example (which also has a relation to very fine-grained parallelism) is provided by superscalar execution and Very Long Instruction Word (VLIW) processing. In superscalar execution, the processor will pick up the instructions and (perhaps dynamically) schedule them. Since superscalar execution can incorporate the simultaneous execution of an integer and floating-point instruction, the compiler can already try to group an integer instruction with a floating-point instruction. With VLIW, the compiler tries to group "normal" instructions into one big instruction, which can be issued in one cycle. This means that the compiler has to determine the parallelism in the code in order to schedule the instructions as densely as possible, into a single "long instruction". This requires a sophisticated compiler employing techniques such as trace scheduling. On the other hand, from a hardware point of view, some microprocessors already implement (or are not far away from that) a so-called Instruction Reorder Buffer, which enables instruction re-ordering. IRB's enable instruction scheduling on a fine granularity basis, and enable out-of-order execution and speculative execution. They smooth out the gaps in the instruction scheduling.

With respect to near-future developments in compiler technology, the trend in hardware (which is support for parallelism) is followed. In general, uncovering parallelism, both single-processor and multi-processor, is the goal for the (near) future. Techniques to establish this include trace scheduling (VLIW execution) and pointer tracking or alias analysis (data dependency analysis), which is particularly important in C and C++. Also important are loop level optimizations, use of compiler directives and memory hierarchy analysis, which aims at optimizing memory-cache traffic and strategies.

We summarize this section on parallel programming models by noting that message-passing implementations are quite important. Also (D)SM machines are providing efficient implementations of message-passing paradigms. High Performance Fortran does not yet seem to be ready for main-stream parallel computing. "Traditional" Fortran compilers continue to include more and more features related to parallelism, also supported by hardware developments. The language of the High Performance Computing community will be Fortran and C, although C++ seems to get more and more attention.

5. Computational Fluid Dynamics Algorithms

In the previous sections we have elaborated on the current situation in High Performance Computing (HPC), with respect to hardware developments (processor, memory), parallel architectures and programming models. We have also tried to forecast trends and expectations in the field of HPC. However, we have not yet discussed the consequences these trends and expectations might have for Computational Fluid Dynamics (CFD). In this section we will briefly try to gain some insight in CFD algorithm development and implementation on current and future architectures.

There are many CFD codes around. In our opinion, from a computational point of view, the greatest common denominator of these CFD codes may be represented by the following set of characteristics:

- Discretization by finite differences, finite elements or finite volumes;
- Many local computations:
    neighbouring elements or volumes

> upwind schemes
> turbulence modelling

- Time stepping: "old" and "new" arrays;
- In general vectorizable to a reasonable extent;
- Parallelization by domain decomposition.

In section 2.2 we have observed that RISC processors will be the most important processor type for the (near) future. However, we have also seen that memory access times will most likely not keep pace with processor speed. In order to efficiently execute codes on RISC processors, it is therefore essential to avoid cache misses. From an implementation point of view, this means the following:

- The processor needs to work as much as possible on the same data: data blocking;
- Use multiple functional units: loop unrolling;
- Increase superscalar execution: loop unrolling.

For algorithm development and implementation, these requirements translate in the need for doing local calculations as much as possible: apply domain decomposition methods not only for parallelization, but also for single (RISC) CPU optimization. For single-CPU optimization, the number of subdomains might be tailored to the size of the data cache.

Domain decomposition methods are also used to parallelize grid-based applications, as CFD applications generally are. For parallel efficiency reasons, the number of subdomains is then tailored to the number of processors of the parallel machine, in order to minimize synchronization and/or communication between processors. To satisfy both the single-CPU and the parallelization requirements on domain decomposition, one could think of a two-stage domain decomposition: "coarse" subdomains for assigning subdomains to processors, and within such a subdomain a further decomposition to utilize the (RISC) processor as efficient as possible.

We want to conclude this section with the observation that DSM architectures are NUMA architectures: Non Uniform Memory Access. Although each processor is able to access the entire memory, access times depend on the physical location of the data. That means that is advantageous, for performance, to have each processor working on a subdomain for which the data are located in memory which is as close to the processor as possible. Hence, just as is the case for true DM machines, paying attention to the actual location of the application's data in memory will improve the application's performance. Compilers for DSM systems implement features to control the physical data location, mostly by compiler directives.

6. Conclusions

In this paper we have foreseen that RISC processors will be the processors used in High Performance Computing. Although RISC processors will still gain some speed, limits will be hit in the next 5 to 10 years. Concerning memory, DRAM access times will remain a bottleneck. Substantial application performance improvements can only be reached by exploiting parallelism. True distributed memory (DM) machines and networks of workstations are curently available and will continue to be available. The traditional drawback of shared memory systems over DM systems (low number of processors) has been removed in distributed shared memory (DSM) systems. Cache coherent DSM machines are already here or have been announced by the most important hardware vendors. DSM machines will most likely dominate the market in the next 5 years. With respect to software, traditional Fortran and message-passing programming paradigms, particularly PVM and MPI, will be the most widely used. Both for single (RISC) CPU performance and for parallelization, domain and data decomposition methods will be heavily used in algorithm development and implementation.

7. Acknowledgement

The author would like to mention that some of the views expressed in this paper have been presented earlier by Greg Astfalk of Convex Computer Corporation. The use of the material contained in [10] is gratefully acknowledged, including comments on a draft version of this paper.

8. References

1. Hockney, R.W. and Jesshope, C.R., *Parallel Computers - Architecture, Programming and Algorithms*, Adam Hilger, Bristol, 1981.

2. Vorst, H.A. van der, *Parallel Rekenen en Supercomputers*, Academic Service, Schoonhoven, Netherlands, 1988.

3. Flynn, M.J., *Some Computer Organizations and their Effectiveness*, IEEE Transactions on Computers, C-21, 1972.

4. Astfalk, G., Brewer, T. and Palmer, G., *Cache Coherency in the Convex MPP*, Convex Computer Corporation, Richardson, TX, USA, February 1994.

5. ANSI/IEEE Standard 1596-1992, *IEEE Standard for Scalable Coherent Interface (SCI)*, IEEE Inc., Piscataway, NJ, USA, 1992.

6. Geist, A., Beguelin, A., Dongarra, J., Jiang, W., Manchek, R. and Sunderam, V., *PVM: Parallel Virtual Machine*, MIT Press, Cambridge, MA, USA, 1994.

7. MPI forum, *MPI: A Message-Passing Interface Standard*, University of Tennessee, Knoxville, TN, USA, 1994.

8. Dongarra, J. and Dunigan, T., *Message-Passing Performance of Various Computers*, Oak Ridge National Laboratory, USA, August 1995.

9. Wylie, B.J.N., Norman, M.G. and Clarke, L.J., *High Performance Fortran: A Perspective*, EPCC, University of Edinburgh, Edinburgh, Scotland, 1992.

10. Astfalk, G., *High Performance Computing and Applied Mathematics: Expectations for the Future*, to appear in Proceedings of ICIAM '95, Zeitschrift fur Angewandte Mathematik und Mechanik (ZAMM), Akademie Verlag, Berlin, Germany, 1996.